FLUSHED

Also by W. Hodding Carter

Stolen Water
Saving the Everglades from Its Friends, Foes, and Florida

A Viking Voyage
In Which an Unlikely Crew of Adventurers Attempts an Epic Journey to the New World

An Illustrated Viking Voyage
Retracing Leif Eriksson's Journey in an Authentic Viking Knarr

Westward Whoa!
In the Wake of Lewis and Clark

FLUSHED

How the Plumber Saved Civilization

W. Hodding Carter

ATRIA BOOKS
New York London Toronto Sydney

ATRIA BOOKS
A Division of Simon & Schuster, Inc.
1230 Avenue of the Americas
New York, NY 10020

First Atria Books trade paperback edition May 2007

ATRIA BOOKS and colophon are registered trademarks of Simon & Schuster, Inc.

For information about special discounts for bulk purchases, please contact Simon & Schuster Special Sales at 1-800-456-6798 or business@simonandschuster.com.

Manufactured in the United States of America

10 9 8 7 6 5 4 3 2 1

The Library of Congress has cataloged the hardcover edition as follows:

Carter, W. Hodding (William Hodding).
Flushed : how the plumber saved civilization / W. Hodding Carter.
p. cm.
Includes bibliographical references.
1. Plumbing—History. 2. Sewerage—History. 3. Water-supply—History. I. Title.

TH6115.C37 2006
696'.109—dc22 2006040703

ISBN-13: 978-0-7434-7408-5
ISBN-10: 0-7434-7408-2
ISBN-13: 978-0-7434-7409-2 (Pbk)
ISBN-10: 0-7434-7409-0 (Pbk)

To my mother, Peggy Carter,
and my father, Hodding Carter,
for potty-training me so well

The society which scorns excellence in plumbing because plumbing is a humble activity, and tolerates shoddiness in philosophy because philosophy is an exalted activity, will have neither good plumbing nor good philosophy. Neither its pipes nor its theories will hold water.

—John W. Gardner
American activist

Contents

FLUSHED

INTRODUCTION

JASMIN

ANGUS, OUR ONE-YEAR-OLD, was busy smearing a mixture of scrambled eggs and peach yogurt in his hair and under his shirt when the phone rang.

"I'll get it!" Anabel, eight, yelled, springing from the chair where she'd been standing on her head—while reading.

"Get your coats on," Lisa countered as she grabbed the phone ahead of Anabel. "The bus'll be here in three minutes. . . . Hello?"

The response from our girls was as prompt and compliant as usual. Anabel went back to doing yoga. Eliza, her twin, stretched across the couch, hand over brow, lamenting her paltry choice in shoes. Helen, six, grabbed Angus by the cheek and growled at him.

"No, nobody's here," Lisa said, laughing. "That's why I'm answering the phone."

"Okay, girls, get going. Now, now," I tried, garnering the same results as Lisa. "RIGHT NOW!"

"You're coming *here* to do this?" she asked the caller, sounding perplexed. "He did, did he?" When Lisa looked over, I immediately knew who was on the phone: Russell. I grabbed the kids and started ushering them toward the door—braving *Aw, Daddy* angry looks, and even a kick in the shin. Anything to get out of the house before she was off the phone.

Lisa hung up as my hand reached the doorknob.

"Hodding."

"Who was it?" I tried.

"You know who it was. I really don't see why people have to come over and use our toilet," she said, helping Helen put on her knapsack. "I don't think I want this to become a common thing."

"Why not? It's a lot of fun."

"It's just not normal. I like Russell, too, but this is really going too far. You just don't invite people over to use the bathroom."

"Dad, you're weird," Helen added, stuffing a Ritz cracker smothered with goat cheese in her mouth. "You really are."

I contemplated explaining to her that normal six-year-olds don't eat chèvre-smeared Ritzes for breakfast, but I was already lost in a plumbing reverie. What's the big deal? I wondered. Why are people so squeamish about relieving their bodies of waste, or, even more to the point, about other people doing it? It's just a normal function we all share in common. Russell may be big and freckly, but what's wrong with inviting him over to try out the new toilet?

Lisa took the girls outside to catch the bus and I tricked Angus into actually eating some of the peach yogurt. When he started to fuss again, I gave him a strip of plastic wrap to distract him.

Lisa returned. "Russell's not here yet?"

"No."

She walked over to the baby. "That's really not good for him."

"He's not holding it in that long," I answered, wondering why she was suddenly so concerned about Russell's bowels.

"No, I mean Angus," she said, and snatched the sheet of plastic wrap off the baby's tray.

Russell stomped in a few minutes later, just after we'd cleaned up.

"I really gotta go," he said, shedding his clogs in obeisance to the growing New England custom of taking one's shoes off before walking through a house.

"Don't you think this is a little weird?" Lisa asked him.

"Hodding asked me to. It wasn't my idea," he responded, sounding a little hurt while walking up the stairs. I'd installed the toilet—actually a high-tech Japanese toilet seat, to be precise—in the bathroom off our bedroom under the premise that the kids wouldn't play with it as much. Wrong premise, of course. They just made it *their* bathroom.

"Well, hold on, I need to get my computer," I said.

"What for?" Russell asked, pausing midflight.

"I've got to capture your experience. It's not every day that somebody uses a Toto Washlet S300, better known as Jasmin, for the first time. The process should be recorded for posterity."

"That's okay, I can remember my own quotes," he said as he continued up the stairs. "Please don't come up."

I grabbed my laptop and sneaked up anyway. The Jasmin is a state-of-the-art washlet—an electronic toilet seat that also functions as a very, very modern bidet, which, incidentally, first came into use in the late 1600s in France. The word *bidet* used to mean "pony." The early bidets consisted of a basin resting on an almost waist-high four-legged stand. You straddled it and washed yourself by hand. The first mechanized bidet did not come on the scene until 1750 and was called a *bidet à seringue*. In this model, a hand pump forced water through a tube located near the bottom of the basin. Perhaps the Jasmin, then, should properly be called a *bidet à computer*.

Jasmin beckons Siren-like on even the coldest days—her ABS plastic has an antibacterial coating and is continually heated at a constant temperature of one hundred degrees. *Sit on me . . . Sit on me, you lovely, filthy human,* she calls temptingly. Of course, you initially hesitate, given that Jasmin, with her gleaming whiteness, looks like a cross between a high-tech electric chair, something out of Woody Allen's *Sleeper,* and a toilet you'd expect to find in a state-of-the-art nursing home, but then you give in, heeding her call. As her caressing warmth envelops your posterior, a motion sensor detects your presence and turns on the deodorizer, a device that, acting like a catalytic converter, actually restructures the stinky molecules into less odorous ones. There is no smell in Jasminland. Simultaneously, the water heater kicks in and warm water trickles through the "wand" that has moved into the ready position. It's almost as if you don't have to do anything at all.

"Everything in life evolves. Including the concept of clean," promised the Toto website, something I happened upon one day while looking for ways to remove chewing gum from my daughter's hair. "What is clean? What's the most effective way of cleaning ourselves?"

Those seemed like good questions. Existential, even. *Stop,* I wanted to say. *You had me at "everything."* But there was more.

"It's the bathroom accessory that turns your bathroom into an oasis of serenity and comfort. The Washlet uses water. Nature's most gentle and essential element. Purifying. Relaxing. Quite simply, the epitome of clean. The evolution of clean."

Just in case a customer wasn't getting the idea, the website also featured a short video that came straight to the point. The washlet, atop a Toto toilet, sat on a pristine rock in a Japanese garden, a soothing waterfall in the background. A male voice-over intoned, "The gentle cleansing features of the washlet not only help keep both the rectal and genital areas fresh and clean, but bring soothing comfort and relief to annoying minor ailments." The scene faded to a bathroom in which a computer-generated, androgynous, see-through human used the washlet, making it very clear where the magic wand sprayed and how versatile it could be. New Age music fluttered in the background and I felt my very soul being massaged. By the end, I found myself agreeing that this was, indeed, "the way nature intended," and suddenly I had to have a very expensive Japanese toilet seat. And so I do.

Now back to Russell.

By the time I reached the bedroom, laptop open and ready, Russell had only just sat down. I knew this by the telltale sound of water trickling into the toilet bowl. The spraying wand was warming up but was still tucked safely under the seat.

At this point, the experience is the same as it has been since the first human found something to sit on while relieving himself. The washlet wouldn't do anything, except purr in anticipation, until Russell was finished. As much as I wanted to ask him how he liked the warm seat—one of my favorite features—I kept quiet, knowing my question would stifle any future spontaneous comments from him.

Minutes passed. Then the water pump kicked in and so I knew Russell was using the remote to control the wand. This was the moment when everyone always said something or at least made a gasp of delight. When Mike Donnelly, our electrician, helped me install the washlet (it needs a GFI outlet), I made him try it out, although he didn't really need to go. Just seconds after I left the room, I heard Mike giggling like a five-year-old. Later he told me, "Hodding, that was one of the most amazing experiences I've ever had." Of course, he might've been employing a bit of exaggeration, but Mike has such a bushy mustache it was hard to read him.

Russell, though, was saying nothing. Not a peep. Not even an *aha.*

I couldn't help myself. Something must be wrong. "You idiot. Did you figure out the controls?" I called through the door, unable to hide my frustration.

"Yeah. . . . What are you doing here?"

"Nothing," I lied. What was with this guy? Where were the exclamations of wonder and delight? No one can resist Jasmin. I sit on her even when I don't have to use the bathroom. It's one of the best experiences *I've* ever had.

The dryer kicked in.

He wasn't doing it right.

"Is that the dryer?"

"Yeah."

"Already?"

"Yeah."

Things went downhill from there. He came out complaining about getting water sprayed all over his pants and described the experience as being, at best, like playing a video game. Clearly, he wasn't ready for the future.

1

The Humble Plumber

The first time I actually took any notice of plumbing, beyond plunging a stopped-up toilet or anointing a sink with Liquid-Plumr, was about nine years ago. My wife and I had bought a fixer-upper in a small town in southern West Virginia. Our contractor had cleaned us out, and so whenever anything went wrong we ended up fixing the problem ourselves. One day I was in the basement trying to shovel out an old slag pile—refuse from when a coal-powered furnace heated the house. (This was West Virginia, after all, where even in 1996 it was almost a sin not to be heating with coal.) It was a thankless job that was going to take a day or two of nonstop shoveling in a five-foot-high space. I'm six-foot-one. Unsurprisingly, after a few hours I grew careless and swung my shovel too far and hit a blindingly white plastic pipe that came from the floor above me and disappeared into the dirt below. The impact cracked the pipe.

I thumped it with my hand a couple of times and determined that it was hollow. Reasoning it wasn't a supply line, I figured I could probably cut out the bad part and put some kind of connector piece in its place. It looked like it might take an hour at most. Who needs a plumber? I thought, laughing confidently to myself.

I measured the pipe, drove to the hardware store, and found out the stuff was a plastic called PVC, short for polyvinyl chloride. Being a youngish homeowner and having never worked at a construction trade, I was completely ignorant of PVC, but it has an overarching history, when you take into consideration that PVC is what we commonly refer to as vinyl and, quite often, simply plastic. The stuff is everywhere, and no wonder, considering what it's made of.

The raw materials for PVC are salt and oil. The passing of electricity through salt water produces chlorine, which is combined with ethylene, derived from oil, to form vinyl chloride. This is then polymerized, which essentially means small molecules of one or more substances are combined to form a larger single macromolecule, or polymer. Additional items, called plasticizers, are added afterward, depending on the desired form or plasticity you want for your PVC. In goes salt and oil and out comes your dashboard.

PVC was invented several times and in different forms starting in the mid-1800s, but it was always too brittle. The inventors did not know how to make it resilient and therefore usable. Finally, in 1926, an American chemist named Waldo Semon, working for B. F. Goodrich, figured out that adding tritolyl phosphate to the mix would make the resulting PVC both resilient and pliable. It was probably one of the most important chemistry breakthroughs of the twentieth century, as far as manufacturing was concerned. Yet still nobody knew what to do with it. Stephen Fenichell, author of the highly readable history *Plastic: The Making of a Synthetic Century,* jokes, "Semon's seminal substance . . . was practically stillborn, since industrial uses for this new water-resistant, fireproof material—which could be molded or extruded into sheets or film—were not immediately apparent." Semon was told to come up with some practical use for PVC or risk losing future research funding. Thus inspired, he cried eureka after watching his wife sew together shower curtains, which at the time were made of rubber-backed cotton. PVC, in the form that would become commonly known as vinyl, was going to replace rubber as the ultimate waterproof material. Thereafter, since it resists both water and to a lesser extent fire, the uses of PVC became seemingly endless, from a coating for electrical wires to a replacement for rubber tires to the irresistible Naugahyde to the ever-present credit card and, of course, to shower curtains.

All great stuff—but there is a huge downside. PVC is toxic to both humans and the environment. When PVC is made, workers, and those living in communities surrounding a PVC plant, are exposed to vinyl chloride, a carcinogen. Chemical companies themselves discovered this when numerous PVC workers from around the world started dying of the same type of liver cancer. Under pressure from the U.S. government, PVC manufacturers began requiring all workers involved in the production of PVC to wear respirators and protective clothing.

The problem with PVC doesn't stop there, though. While in use, some toxic additives can leach out. Phthalates, plasticizers found in vinyl flooring and in plastics used in car interiors, have been found to cause cancer and to lead to demasculization of male fetuses. That new car smell we all love so much? That's the volatization of phthalates caused by the hot sun.

PVC also wreaks havoc at the end of its life. At recycling centers, when it inadvertently gets tossed in with recyclable plastics, it burns and destroys the surrounding plastics and harms the recycling equipment. Its burning also frees those same dioxins released in its manufacture.

Regardless, the world currently produces about eighteen million tons of PVC a year and it is probably the most lucrative chemical compound in the world. As far as plumbing goes, since the 1950s PVC has slowly become the material of choice for domestic drainage and ventilation. It's lighter and cheaper and easier to use than either cast iron or terra-cotta. You could conceivably carry an entire house-worth of PVC waste plumbing on your back. You could cut it to the right size and, after gluing (which is also toxic), be finished in two days. In other words, PVC is here to stay.

Back at my hardware store in West Virginia, after I described the situation in my basement, the salesman agreed that I had probably hit a drainpipe and that given its size—four-inch diameter—it was probably the main drain, or waste stack. Although it serviced two toilets, three sinks, a shower, a bath, and the washing machine, he agreed with me that it didn't seem like a big problem. My eight-months-pregnant-with-twins wife was at work, so I didn't have to worry about anyone using the water and having anything leak all over the basement. I'd only busted a short section. He guessed the repair wouldn't even take an hour. I bought everything he suggested: pipe glue, pipe cleaner, an extra length of PVC pipe, and three couplings with female openings. Presumably, I'd only need two; the third was backup, a required item, I'd soon learn, for all plumbing situations.

I returned to the basement almost giddy. I was doing home repair and plumbing at the same time. Take that, Bob Vila. I'd once claimed on a Peace Corps application that I'd make a great water works volunteer because I'd tinkered with my family's toilets since I was a kid (only true to the extent that I had reset the lift chain and plunged the bowl countless times), but the Peace Corps wisely put me to work teaching history in a secondary school. Here, finally, was my chance to prove my claims.

I began by cutting out the busted part using an electric reciprocating saw. When finished, I had a foot-long gap that I could now fill with replacement pipe and a couple of couplings. Victory. For about five seconds. Then I noticed that the bottom part of the pipe was movable. I gingerly pulled on it and it popped out like a greased finger from a glass Coke bottle.

I was staring into a hole in my basement floor and the malodorous gas I'd begun noticing while cutting the pipe was suddenly inescapable. I was sucking in hydrogen sulfide, methane, and other substances from my entire neighborhood's sewage. Sewer gas, because of the hydrogen sulfide, is as toxic as cyanide. At low dosage levels, it causes eye irritation, dizziness, and headaches—all of which I was feeling. At higher doses, you lose your sense of smell, which often leads to fatal doses because you don't know you're breathing the bad stuff anymore. You breath in a few more times, it paralyzes your lungs, and you die. You really don't want to be around sewer gas in enclosed, low-ceilinged spaces (like a musty old basement) because the gas is heavier than air. Luckily, I realized what was befalling me and I stuffed a rag in the gaping pipe. In the early days of mass municipal plumbing, sewer lines often had backflow flaps that would stop not only other people's sewage from coming up your sewer pipe but also their gases. In an 1888 plumber's catalog, the Penn Company claimed of its "patented improved sewer pipe trap" that "backwater and draughts of air that frequently force poisonous gases and the contents of sewers into houses cannot pass this trap." However, the opposite was usually true; the flaps hardly ever worked, invariably causing clogs and failing altogether after relatively short time periods.

I grabbed the pick and began swinging for all I was

worth—in my now-customary crouch on account of the basement's five-foot height. I had to be finished in four hours before Lisa returned from work.

About two hours later, I had chipped away enough of the floor to discover that the pipe made a ninety-degree turn about a foot below the basement floor. After the turn, though, the pipe wasn't plastic anymore. It wasn't even metal. It was terra-cotta. And it was cracked. From the looks of it, it had been cracked for quite some time—the dirt surrounding it was moist and stinky. Our waste was intermingled with that of the previous owners, heralding a new beginning. Life arises from shit, the grand enricher. (I think the fumes were still getting to me.)

It was hard to know for certain how cracked it really was because to unearth the pipe I had to strike all around it with my pick. For all I knew, I could have been cracking it myself. I wouldn't be done before Lisa got home.

Two days later, I had uncovered only six feet of broken pipe. The dirt floor had switched to concrete after a few feet and so I'd also had a day of jackhammering. However, I'd finally come to a place where the pipe looked strong and solid. I didn't know this for sure, but then I also didn't plan on living in this house for the rest of my life. I stopped digging.

It'd been a rough two days, but I remember lying on my back in the muck, staring up at the plastic and metal pipes that various people had installed, and marveling at my plumbing's overall structure. It looked so simple and graceful—maybe not easy to do, but simple in an artistic way. It was a labyrinth of pipes that made sense to somebody, and as I stared at it, following each pipe to its beginning and terminus, it began to make some sense to me, too. It was no longer inexplicable and out of bounds.

Here, finally, by literally digging in my own and others' waste, I finally understood something about a house. It was such a cathartic experience that I decided then and there that someday I would become a plumber. That night, after I had successfully fitted PVC to clay piping using a rubber sleeve with hose clamps, I told my wife about my epiphany. She'd just come home from work and had a pained look on her face. Upon my proclamation, she merely nodded and asked if it was now okay to use the bathroom. Being pregnant with twins had made the last few days a bit stressful for her.

Almost ten years have passed since that fateful moment, and while I haven't become a plumber, I want to know what a plumber is, what he was, and how he got us to where we are today. Obviously, he's the guy who works on water systems flowing into, out of, and within buildings. But can't we also say he's the guy who first created urban water systems, like Rome's famed aqueducts and the in-some-circles well-known Cloaca Maxima? (Cloaca Maxima, which means "the greatest sewer," even had a goddess, Cloacina, watching over it and was built around 500 BC—predating the first water-supply aqueduct by two hundred years—to improve drainage of the city of Rome. An embankment envelops the Cloaca today but it still performs some drainage. You might have stared at it admiringly without even knowing it: John Singer Sargent made a mesmerizing watercolor of a woman washing clothes at its outfall in 1869.) Yes, those guys were called *aquarii* and were engineers of a kind, but at heart weren't they really plumbers? Haven't we overlooked the contributions of this profession for far too long?

And then more questions arise. When I tell friends and strangers alike about my interest, they invariably smirk and/or look away. Why is this a difficult subject? What made bodily functions so private? They weren't always so, and in some parts of the world still aren't even today. And what makes bathroom humor so popular? Does it mark the downfall of a society, as some pundits would have us believe, or is it the dawn of a new era? Some historians have suggested that the use of silver and gold chamber pots by the Romans marked the decline of that civilization. Does breaking taboos surrounding excretion amount to the same thing? Is our end nigh?

In answering these questions and more, I hope to show this subject in a new light and perhaps convince a few readers to genuflect before the porcelain god even when not overcoming the effects of overindulgence. At the very least, if you're opposed to revering common idols, please pause and consider the men and women who have brought us our water for thousands of years. You wouldn't be here if it weren't for them.

One of the world's earliest known civilizations, the Harappan of the Indus Valley, circa 3000 BC, is noted for one professional's work above all else. Today's scholars marvel at his ingenuity and precocious skills. Whenever, wherever a society arose from this point on, it was on this humble professional's back. The sparkling Athenians let him create unequaled contraptions, while the austere Spartans wasted their time throwing spears and performing feats of strength and agility. The Romans worshipped his complex constructions, placing a goddess in charge of his effluent, and gave him the name by which we still know him today.

European monks selfishly let him work his wonders in their monasteries while their neighbors wallowed in filth and disease. The British Empire awarded him medals of honor for his designs while the Americans and French played catch-up, always imitating but never quite equaling the British professional's work.

This unsung hero of human history was, of course, the Brain of Drains, the Hub of Tubs, the Power of Showers, the Brewer of Sewers . . . the humble plumber.

The Irish may have saved civilization, once, but plumbers have done so countless times.

Admittedly, I might resort to a bit of hyperbole when considering the contributions of this profession. Many in my family think it's because the subject affords me the opportunity to talk about things scatological under the guise of a more serious topic, but they're wholly mistaken. I like plumbing—the art, the science, the craft, and the history—for the simple reason that it is so undeniably important and necessary. Without it, even with electricity and the Internet, we'd be living in a state akin to that found 150 years ago (and there are some two billion people still living in such a state today). Yet we take it for granted. We don't appreciate it. We never marvel at it. We never hold forth on why plumbing is such an important utility and why it has always been what separates the developed from the undeveloped—from 5,000 years ago to today.

Let there be no mistake. A clean modern water supply, working toilets, and environmentally safe sewage systems are what divide the successful from the unsuccessful, the comfortable from the uncomfortable, and the privileged from the unprivileged. And it's been this way for thousands of years.

Think about it. On a fundamental level, what is more important to you: Electricity or running water? Your computer or your toilet? Shock and Awe or a sewer system?

Without electric lights, you could use a candle. Without water coming in or going out, however, where are you going to go? It'd be a bad enough situation for the guy living in rural Idaho suddenly having to run outside, dig a hole, and cover it back up each time, but what about those of you in New York City?

A lot of things would go awry without proper plumbing. We would not have our much-vaunted general good health, and without our health, we would not have the time or energy to learn and prosper. Things didn't start getting better for Europeans in the eighteenth century because of improved health care; on the contrary, doctors back then probably killed more patients than they saved. People started living a little longer and having a bit more freedom because cities started to supply their residents with water and then, very slowly in places like London and Paris, figured out how to get rid of waste.

Unknowingly, we owe our modern way of life to Roman lead workers, to the unnamed craftsmen who plumbed early European cities, and to the ingenious work of such men as Sir John Harington, inventor of the water closet in 1596, and Thomas Twyford, champion of the one-piece toilet. Thomas Crapper didn't invent the toilet (a widely held misconception), but he did improve it, marketing a valveless toilet in the late 1800s in England. Crapper was a simple plumber and manufacturer, but we know his name so well because his toilet was the most popular brand in England during the First World War. Our doughboys, it is guessed, would announce they were going to the "Crapper." The euphemism crossed the ocean when they came home. By some strange coincidence, *crap,* as slang for feces, was used long before *Crapper* came on the scene. Etymologists believe it is derived from the Dutch word *krappen,* which means to cut off, and its earliest known written usage was in 1846, when Thomas Crapper was only ten years old. This is a perfect example of an aptronym, a name that is suited to the profession of its owner.

Plumbers and inventors like these were once held in such esteem that they were even the envy of royalty. When Queen Victoria's son Edward, the Prince of Wales, suffered from a near-deadly case of typhoid, a plumber realized that a faulty water closet in the house where the prince had been staying was breeding typhus and fixed the situation. After he recovered, Prince Edward declared, "If I could not be a prince, I would rather be a plumber."

Today, though, the plumber is, more often than not, merely the butt of jokes. Plumber's crack is about all many of us know of this age-old profession. Yet, clearly, plumbing waters run much deeper. (By the way, a Web-based suspender company recently devised a strapping device that they claim will end plumber's crack forever. The device is being product-tested at this very moment. Early word has it that plumber's crack could soon be a thing of the past.)

So, here's your chance to catch up. See how plumbing's conveniences have flowed through time and shaped our modern world. It's one of those things we all have in common, poor and rich alike: our need for and reliance on plumbing. Plumbing is the great equalizer. After all, as the title of a popular children's book from Japan points out, *Everyone Poops.*

2

When Lead Ruled

ALTHOUGH WE THINK of it as a modern convenience, plumbing, as in a conduit connected to a building for the purpose of bringing in water and taking out used water, has been around for a long, long time. At least seven thousand years and probably much longer. But, regrettably, it hasn't been a straight, interconnected march of progress through time. In other words, it's not as if some genius made the first terra-cotta pipes, connected them, brought water to his master's house, and then the invention improved from then to now. Instead, plumbing has come in fits and starts.

The Chinese probably practiced the earliest known form of plumbing in the fifth millennium BC. They used pipes made from bamboo. Cool stuff, but that's about all we know. Did these pipes enter private homes? Feed a town cistern? It's anybody's guess, but nevertheless, they did use pipes before anybody else. Then sometime in the third millennium BC, the previously mentioned Harappan civilization of the Indus Valley built water systems that in many ways would rival and surpass any other water system, except that of the Romans, until the middle of the nineteenth century.

In Mohenjodaro, a major Harappan city in what is now Pakistan, not only did the ruling elite have indoor plumbing, but so did everybody else living within the town walls. Each solid brick home, many of which were two-storied, contained a bathing room with an adult-length bath made of tightly packed bricks that were sealed and covered with a gypsum plaster, making the bathing areas virtually waterproof. They also had the first known method for allowing humans to excrete their waste in their homes and have it washed out of the building—a simple hole built into the floor that dropped down into the same brick drain that took out the bathwater. From there the wastewater went into a cesspit that was designed to empty into covered sewer mains as it filled up. These mains were tall enough for people to walk through, presumably to facilitate repair work. The sewer in turn emptied outside the city walls into the local river. Nothing like this existed anywhere else in the world. Actually, even today, nothing like this exists for nearly half the world's population.

Water usage continued to develop in the East with the building of water systems called *qanats* by the Persians—elaborate tunnels that sometimes transported water for miles and miles to walled cities. The beginning of the qanat always started higher in elevation than the terminus, since it ran solely by gravity. Slaves dug the qanat tunnels by hand and made them no bigger around than an average man's girth. The tunnels were usually unlined, but in cases where they passed through loose soil, rings of clay reinforced the sides of the walls. The workers also dug vertical shafts every twenty-five yards or so to provide ventilation and access. The vertical shafts are thought to have helped in sighting the tunnels and setting the proper drop in elevation. Qanats were constructed without any surveying tools. (Iranian qanat builders of the twentieth century kept their tunnels straight from shaft to shaft by keeping two lit lanterns placed a few feet apart directly behind the digger. When the digger looked back and saw that the one light was superimposed on the other, he knew he was going straight.)

In most cases the qanats branched off into aboveground canals that irrigated local crops. However, sometimes they fed an underground cistern, usually dug out of rock, that supplied drinking water in the dry environment where most of the qanats were built. Baths have been found at a number of archaeological sites throughout the Middle East, also indicating that the qanats were used for more than agricultural purposes.

Over time, the qanats spread throughout the East, into western China and much of the Mediterranean region. In North Africa, ancient qanats still deliver water to villages.

Meanwhile, in areas where there was plenty of rainwater, plumbing advanced far beyond simple water delivery. Around 2000 BC a Minoan ruler, perhaps King Minos of the Daedalus and Icarus myth, built a formidable castle covering five and a half acres in Knossos, Crete, complete with running water and even a water closet, meaning a room for defecating and urinating, equipped with a wooden seat. Captured rainwater, held in cisterns, cascaded through the walls in ingenious tapering terra-cotta pipes. The outsides of the pipes had handle-shaped rings that allowed the pipes to be hitched tightly together, perhaps with leather string. Once snug, the joints were bonded with cement (Greeks and Romans were the first to use a limestone-based cement as mortar; the Egyptians appeared to use it for plastering). These pipes were engineering marvels on the inside as well; the slightly angled connections created a flow that discouraged sedimentation.

Besides flushing out the water closet, the pipes fed various cisterns and baths, one of which is now known as the queen's bath. It was a large room equipped with a terra-cotta tub adorned on the inside with a mesmerizing ring of painted reeds. A frieze of waves rolled high above the entire room. In the next room, maybe the queen's dressing room, dolphins adorned part of the walls, dashing over two doorways, one of which apparently entered the queen's water closet. Drains from the bath and water closets emptied into a main drain that was stone-lined and tall enough for a man to walk through, much like the sewers found in Mohenjodaro.

King Minos's plumbing represented a luxurious breakthrough, but he wasn't much of a trendsetter; no evidence of similar plumbing around the Mediterranean from this time exists. In fact, plumbing technology didn't advance for the next thousand years or so, until the Greeks started making pipes out of lead. Lead had been in use since 8000 BC—ever since somebody discovered that the malleable substance that sloughs off silver ore during refining was useful—but no one had been making pipes with it.

Why didn't people start making lead pipes as soon as they were making terra-cotta ones? History isn't very clear on this—plumbing not being one of those subjects academics scrutinize and fret over—but I'll hazard a semi-educated guess. Although lead is malleable, it's still not as simple to shape into a hollow cylinder as you might think. My own Neanderthalish attempts at pipe making confirmed this—but more on that later. A few people obviously tried to do so—there's scattered evidence of lead pipes in use some three thousand years ago—but it wasn't yet practical. What it required was more leisure time, more time to think, and more important, more time to think of ways to be pampered.

The Greeks had such leisure time, along with the need for improved water supply and the necessary natural resources in water and lead. First, though, they one-upped the qanats by making the first aqueducts—a water delivery system powered by gravity and delivered through a covered (usually) conduit made of stone or bricks (lined with cement or some other type of plaster), lead, or terra-cotta. The aqueducts also differed from qanats in that they were not for agriculture but for fountains and baths.

The Greek aqueducts did not carry the same volume of water as the later, more famous Roman aqueducts because they typically used small terra-cotta pipes of only eight to ten inches in diameter, as compared to the two-foot-wide lined channels that comprised the Roman systems. But they were still marvelous feats of engineering. The aqueducts usually started at a stream in the mountains where the water was pure. The workers would construct a reservoir, lining it with stones and other material, and then attach a pipe to the reservoir. From there, it was a simple matter of connecting pipe after pipe, sealing the joints with cement. Sometimes these pipes were laid in tunnels that went through mountains, and other times they simply ran slightly belowground in a channel covered over with brick or stone.

While early Greek societies did have aqueducts of forty miles in length and fountains flowing with unprecedented amounts of water, it took the Hellenistic Greeks (Helen of Troy era, 323 to 150 BC) to really advance things by shortening many of the old aqueducts and by using more and more lead. During this era, the supply lines in most communities went from consisting of mainly terra-cotta to lead. If there was a fountain, which any decent town had, it was surely fed by lead pipes.

They shortened the old terra-cotta pipelines by constructing something called inverted siphons to cross valleys. Since water will return to its own level, the Greeks realized they could build a pipeline that went down an escarpment, ran along a valley, and then came back up to near the same level as the previous escarpment and the water would continue to flow to the city farther below. In other words, they didn't have to build a bridge for the water system. On a small scale, these U-shaped dips could be made with the traditional terra-cotta pipes. In more extreme situations they had to make pipes out of stone and even lead. One particularly impressive siphon consisted of two miles of lead piping. Each of these pipes was about five feet in length, weighed approximately twenty-five hundred pounds, and was strong enough to hold up to the extreme pressures of the siphon.

While such leadworks were impressive, as was the resulting abundance of water in Greek communities, it is the Roman word for plumber, *plumbarius,* that has survived the ages; and it is the Romans who made plumbing a profession resembling that of modern times, thanks to their outright obsession with the building of their baths; and it is the Romans whom we think of as the first true master plumbers. Wherever they went, from Africa to England, they built overflowing aqueducts and sumptuous baths. There's been nothing like it before or since—this religious obeisance to running water. You see, that is what it was about, not health or cleanliness, not just water for the masses, but instead, the power of running water. Running water was the unique, clarifying symbol of the wealth and power of the Roman Empire. The empire could tap water fifty miles away, bring it to the city, and then just let it keep on flowing. The Romans didn't bother with reservoirs or any type of system that allowed the water to be preserved. They simply let it flow—and in some cases for thousands of years, if you consider that a few of their aqueducts are still in service today.

And where there were Roman baths and aqueducts, there were pipes. Most people marvel at the above-ground wonders like France's Roman aqueduct, Pont du Gard, that spans the Gardon River at a height of 150 feet, or the seemingly never-ending arcade of Rome's Claudia Aqueduct, but pipes are what really set the Romans apart. The aqueducts themselves were simply improvements on the qanats and the Greek aqueducts; for the most part, Roman aqueducts were built using these two predecessors' principles. It was the lead pipes that lifted Roman plumbing above the rest.

Once the cool, rushing mountain water entered the city into a collection area called a *castellum*—an enormous cistern made of cement-lined stone—the *plumbarii* took over. From that point on, lead ruled.

Water left the *castellum* via lead pipes that fed

smaller water-containment areas within the city walls. From those smaller *castellum,* smaller lead pipes fed different bathhouses and the public fountains, and even smaller lead pipes ran to private homes. The outlet for water going to private homes was placed highest up in the *castellum,* then the came the bathhouses' outlet, and at the bottom was the outlet for the public fountains, where most citizens gathered their water. The outlets were placed in this order so that the most important destination, the fountains, ran out of water last in dry times.

Although most homes did not have their own water supply, those lucky enough to have indoor plumbing were charged an annual fee based on the size of their pipes. The typical supply line for a home was a *quinaria*-sized pipe.* (The largest pipe, one hundred *quinaria,* was used for supplying public bathhouses.) The fee was based on pipe size because, as previously mentioned, the water was never turned off. Water flowed day and night, seven days a week. Shutoff valves weren't used, since stopping the water would cause the system to overflow. The only way to know

*Scholars do not agree on what this actually means. The definition I like best is the one that suggests that a *quinaria* is equal to five digits and that a digit is the length of a finger's circumference. A measurement of five digits, however, refers to the width of the sheet of lead from which the pipe was made. In other words, a *quinaria* was five digits wide before being shaped into a pipe.

how much somebody was consuming was to know his pipe size. However, tricky homeowners hired plumbers to put in larger pipes when nobody was looking, especially when they had their own private baths.

The public baths, though, were what it was all about. More sybaritic than modern-day Turkish and Russian baths, the Roman baths put places like today's Greenbrier spa in White Sulphur Springs, West Virginia, to shame. Olympiodorus, a widely traveled Greek poet known for his factual observations, says the Baths of Antoninus in Carthage, the largest ever built in the Roman Empire, contained sixteen hundred marble seats for the latrines alone. A single bathhouse probably had more marble in it than all of Saddam Hussein's palaces put together.

The baths presented not only a spot to unwind and wash, but also to conduct business, and one of the favorite areas for doing so was the latrines. Ancient Romans weren't shy about expelling bodily wastes, so the latrines were set up for conversation, often in a wide-open room with no partitions separating the occupants. Water flowed from cisterns in the building via either lead pipes or open channels, and upon entering the latrine room a channel flowed in front of the row of latrines and then around underneath the latrine to catch what fell from the occupant. The user would dip a sponge attached to a stick into the water flowing in the open channel in front of the latrine seats and use that to wipe.

The rest of the facility was a marvel of Roman technology as well. Backing up a bit, we see that a typical bath had a large cistern—really large, like a uniformly deep Olympic-sized pool—adjacent to the main building. The cistern was compartmentalized so that one area could be emptied out for cleaning and maintenance while the other still held water. (The main problems were sediment and a buildup of calcium on the pipes.) From the cistern, lead pipes carried water in underground tunnels to the bathhouse proper. The tunnels made repairs easier to accomplish.

There were usually three different types of rooms available in the bath: the *caldarium,* the *tepidarium,* and the *frigidarium*. Plumbers piped the cold water directly from the cistern to the *frigidarium,* the cold room. The water for the other two rooms was heated by wood-fired furnaces located in the *hypocaust*—the heating area. The *hypocaust* took up the entire basement, which allowed heated air to flow beneath the *caldarium* floor, and other rooms as well, and inside the walls through many terra-cotta pipes. This was an early form of radiant heat that we're only now beginning to emulate in the United States with the hot-water radiant heating systems that are spreading across the northern states.

Using the baths was a splendid, multi-hour affair. Say you've just spent a tough day watching your favorite gladiator get eaten by a lion. To lighten your load, you head to the closest bathhouse, which is reasonably priced, since the idea is to provide a place for all free Romans (men and women used the baths separately until the empire's later years) to relax and get clean. So, even if you've lost a huge sum betting on that gladiator, you can still afford your bath.

A smile crosses your face as you glimpse the towering complex from down the street with its row of marble pillars and mural-painted walls. Knowing that you will soon use the bath's latrines and maybe even pick up a bit of gossip, you pass up the free piss pots set outside fullers' shops. (Fullers—meaning a person who cleans, thickens, and/or felts cloth—used urine for its ammonia, which makes a great dye and cloth-stiffener.)

You climb the stairs, cross the threshold, and are greeted with a comforting blast of warm, thyme-scented air. Everybody who is anybody gets rubbed and cleansed with olive oil infused with thyme. It's really the only way to get clean. Taking a deep breath, you're already feeling much calmer, and you haven't even undressed. After paying the small fee (something akin to the price of renting a DVD today), you enter the *apodyterium,* the dressing room, where you disrobe, put your clothes on a shelf or in a small cubbylike locker, and head off to the *palestra,* gymnasium, wearing a light *subligaculum,* undergarment—only the silly Greeks exercise naked. There, you might lift barbells, wrestle, or play a game of ball. If not in a sporty mood, you go straight to the massage rooms located on the side of the gym. Here you get massaged with the thyme-infused oil and have the day's dirt scraped off you with a curved instrument called a *strigil.* Once scraped raw and now tingling with excitement, you go to the *caldarium* for a sauna or steam bath trailed by slaves carrying towels, oil, and perhaps a *strigil*—the more scraping, the better. The *caldarium* has a very hot pool in the center that you sit in, or beside, to sweat and soak up aromatic steam. If this isn't hot enough for you, you can dip into the adjacent steam and sauna rooms.

Next you take a quick plunge in the cold pools of the *frigidarium* or, if feeling weak of heart, cool off more gradually in the *tepidarium.* On a contemplative day, as you wander back and forth between the various rooms, you might marvel at the beauty of the building, with its marble walls and floors, vaulted ceilings, and endless mosaics, frescoes, statues, and paintings. Or maybe not, since you come here nearly every day of the week.

Once done bathing, you dress and stroll the flower-strewn grounds, buy something to eat in your favorite cantina located in the bath premises, or head off to one of the attached brothels. Many of the men's changing rooms have racy mosaics and paintings adorning the walls, perhaps as a way of drumming up business for the in-house brothels. In Pompeii, in one bathhouse, each cubby is painted with a different sex scene so that even the illiterate can remember where their clothes are. Message to self: remember, upside-down contortion with painful smile, three on one.

Whether a food stall or a brothel is your last stop, a few hours later you go home a contented Roman citizen, knowing full well why the Roman Empire has lasted as long as it has. This is the way life is meant to be.

Hmm?
Life of slave?
% of free Romans
to slaves?

3

Do As the Romans Do

Ever wondered what it was like to be a Roman plumber?

Well, think twice about it. Getting too close to this subject can be quite unsettling—as I learned the hard way. It's no wonder Pliny the Elder, *the* authority on everything in the Roman world and author of *Natural History,* a twelve-volumed tome on everything in the known world, remained completely mum on the business of plumbing.

Yet, although quiet about how plumbers plumbed, Pliny did warn of lead: "While it is being melted, the breathing passages should be protected during the operation, otherwise the noxious and deadly vapor of the lead furnace is inhaled."

Alas, I didn't see that line until much too late.

I spent three days in a row working with molten lead in an attempt to be a Roman *plumbarius,* and afterward I passed another three days staring, unable to do much, and feeling listless and unfocused—pretty normal activities for me, really, except that I also felt weak and had an upset stomach. Lisa kept asking me what was wrong, was I having an affair, did I need to see a therapist? I'd shake my head like I was trying to recover from a concussion and respond as from within a tunnel, "No, no, I'm fine. I think I just have the flu."

"Think it has something to do with your playacting?"

"Huh?"

"The Roman plumber stuff?"

Finally, convinced it might not be the flu and to prove it wasn't an affair, I had blood drawn to see if my lead levels were too high.

As luck would have it, the lab messed up my blood sample—something about it not coagulating properly. My doctor, though, was convinced that I had indeed poisoned myself but reassured me that while lead fumes affect you faster and more demonstrably than lead solids, they also leave your system quickly. I think he also said something about the effects not being lasting, but I can't remember.

I got the idea to make my own lead pipes while reading *Roman Aqueducts and Water Supply* by A. Trevor Hodge, a friendly old classicist at Carleton University way up in Ottawa. I know he's friendly because I called him the morning I finished his five-hundred-page book, having found a phone number for him on the Internet.

His wife answered the phone. Evidently, I'd called him at home.

"I just finished reading *Roman Aqueducts and Water Supply*. What a great book, Professor Hodge," I said, thrown off by the fact that he'd actually taken the call. I'd been expecting it to be his office number and that an assistant would take a message, thus giving me time to order my thoughts. "It's really a fun read."

"Well. I'm certainly glad you called, then," he answered, sounding amused but also slightly cautious. Perhaps this didn't happen every day.

I was at a loss. Why had I called? Oh, yeah. "I was wondering how did the plumbers connect the pipes *in situ*?" I actually said "*in situ*." This was a professor, after all. "How could they heat the pipes properly to create enough heat to make the solder work?"

"You know, I've often wondered that myself and no one seems to have a definitive answer. It's not something they write about." "They" being the archaeologists and academics who study Roman artifacts.

"Well, I'm thinking about making some pipes myself and trying out different methods."

"Oh, good. That's really the only way, isn't it?" One of the cool things about his book, as compared to most academics' works, is that he actually tried out his theories; in writing about Roman surveying tools, he made some himself to make sure they worked the way other scholars said they did.

There then occurred what became an increasingly long pause in our conversation. I realized that I didn't have anything else to say. Here he'd written a meticulously researched and gracefully nuanced study, and all I wanted to know was how the plumbers did their job. There was a world of other things a person could ask or say to him, like, *The sighting of those serpentine aqueducts sure was something,* if such things were to come to mind. But, being plumber-obsessed, I didn't really care about the aqueducts, beyond their being vehicles for the plumbers to get to work.

"Well, then, I guess we better . . ." he said.

"Right, right. Sorry. I really just want to tell you again what a wonderful book you wrote. . . ."

"Thank you. Good luck with the pipes, and if you have any more questions just ring again," he said cheerily, and then there was another pause. "By the way, how did you get this number?"

"It was on the university's website. It's your home number, isn't it? Sorry about that."

"Not a problem," he said, but I felt him making a mental note to contact the university. I could have been some plumbing-fixated wacko. "Nice speaking with you."

Inspired and feeling that I didn't want to have called him for nothing, I immediately set out to make my very own Roman water pipes and solve this mystery for both of us and aid the advancement of academic knowledge. *Hoddingus Plumbarius,* to the *rescueus!*

Here's the question/problem: How do you connect one lead pipe to another when it's in a trench and you don't have blowtorches or any other form of maneuverable heat source hot enough to melt lead or tin? And, on a simpler note, just how hard was it to make lead pipes by hand that would be durable enough to hold pressurized water? It was lucky I had this second question in mind as well because, as you will see, I never got around to answering the first.

His book made the answer to the latter question seem simple enough: "The process began by pouring out from the melting pot on to a flat surface enough lead to make a section of pipe of the desired gauge." (I'd assumed the Romans cast them, but the first metal pipes weren't cast until the thirteenth century.)

Okay, melting the lead would be easy, but wouldn't handling molten lead be a tad dangerous? Undeterred, I read on: "What the surface usually was is unknown. It may have been stone, more probably sand or clay, smoothed flat, with slats of wood of the appropriate thickness to form the sides of the mould and stop the lead spreading. The length was uniform, for all pipes came in lengths of ten Roman feet." Once poured, the freshly minted lead sheets were simply curled around a rod of uniform diameter to shape the metal into pipes.

Sounds easy, right? I certainly thought it was going to be. In fact, I thought nothing of this curling-sheets-of-lead process and, as can be seen by my questions for Dr. Hodge, was only concerned with how to connect the finished product. That was my first mistake.

The second was my mold. Since nobody knows what the mold's surface was, and not having access to a ten-foot slab of marble that I could pour molten lead onto, I decided to make my entire mold out of wood, not just the sides.

I cut out a narrow five-digit-by-ten-foot length of particle board and framed it in with the straightest two-by-fours I could find. Others have figured out that five digits is equal to about three and a half inches, so this part went well, and it ended up looking pretty cool to me.

I called up our only local foundry to see if I could melt and pour the lead there, since Lisa didn't think I should do it in our kitchen.

"You've already made the mold, huh?" Richard Remson, owner of the foundry, asked. "I really don't think they would have used wood. The melting point of lead is about eight hundred degrees. Wood burns at a much lower temperature."

"How much lower?"

"Paper burns at four fifty-one, right?" he asked, with a slight, almost imperceptible, chuckle.

"Oh, right. *Fahrenheit 451.* Very good." I tried to laugh, realizing for the first time that I was in way over my head. He suggested I cake it with two layers of joint compound (that's the stuff used to fill in the seams on drywall), then scorch the whole thing with a blowtorch.

"You can't have any moisture. Molten lead recoils from moisture and it'll end up flying back onto you. Not a good thing." And to think I almost didn't call him.

So for two days I smoothed joint compound in the frame, faintly cursing Richard, whom I didn't even know, because I thought the plaster was ruining my smooth finish. I eventually stopped cursing him, through, when I realized the stuff was filling in the myriad of cracks and openings scattered throughout the poorly made frame. Also, I was able to sand it so that it was smoother than my original finish.

I just needed some lead, which, given the current hysteria regarding this natural compound, isn't so easy to come across in large quantities. You can still get lead sinkers for fishing, but that would be too expensive. My local hobby shop doesn't carry lead and a friend of mine who owns a fancy house-flashing company told me he doesn't use lead very often, although later his wife told me he uses it practically every day; clearly, long-term exposure does have its drawbacks. Then I remembered the flashing that lines the eyebrows of all the windows of my house. It's made of lead. I'd buy eighty pounds of it (according to Frontinus, the *curator aquarum* under Emperor Vespasian and author of a study of the Roman water system called *De Aquis Urbis Romae,* a ten-foot *quinaria* pipe weighed the approximate equivalent of forty pounds) so I could melt it down to make two pipes. I couldn't just use the flashing itself for my pipes. Not only would that be cheating, but at one centimeter in thickness, it was way too thin. I was looking to make pipes with about a half-inch thickness.

The metal's density impressed me when I carried a roll of it from the back of the hardware store to the front. While I could bend these sheets of flashing with just two fingers, a six-inch by twelve-foot roll was surprisingly heavy. (Gold, I'm told, is twice as heavy and just as supple, but I have yet to have any personal experience with forty-pound sheets of gold. Maybe someday.) I finally understood the genesis of the lead line, a hunk of lead attached to a line marked by fathoms used for making soundings aboard boats—and also the plumb bob, a pointed weight of lead attached to a string to get a straight vertical line when building. Nothing like cheap, heavy metal to weigh you down. Also, things dawning on me quickly as I toted my lead around, I finally realized why the symbol for lead is Pb. The Latin word for lead is *plumbum.* It is also where we get the expression "plumb the depths" and maybe even "he's plumb crazy," as in he's wacko from lead. And soon enough, I would have an even greater appreciation for the word *plumbism,* meaning lead poisoning.

At first I was very careful, wearing work gloves whenever I carried the lead (except at the hardware shop, where I didn't want to appear a weenie). When I brought it over to Richard's I even packed three sets of latex gloves. Then I saw his shop: blowtorches, shards of multicolored glass spilling out of barrels, piles of hardened overflowed metals, toxic substances everywhere, and, well, I just decided to blend in with the environment and work with my bare hands. That was probably yet another mistake.

Richard, a tall, laconic graduate of the Rhode Island School of Design, fired up the furnace and directed me to load half the lead into a large steel ladle. The ladle, with its four-foot handle and a round bowl half a foot wide, looked like a relic from the Middle Ages. While the lead was melting and heating to eight-hundred degrees, Richard scorched the inside of my frame, actually setting it on fire in a few places, to evaporate any remaining moisture and give the wood a higher burning temp (it takes more heat to combust something already burned).

"I still don't think this is what they would have used," he said, staring down at my creation. It looked like a charred boa constrictor coffin. "They were doing high-end production for their day—almost mass production. Wood just wouldn't have held up." He looked up at me and asked almost as an afterthought, "Is it level?"

"Um, yeah, I think so."

Suddenly it was like I was waking from a dream. This was molten metal we were about to fool around with. It was going to find its own level and I had only eyeballed the two-by-fours to check if they were straight. Not once had I thought to use a level. The boards were probably as warped as my brain. And, come to think of it, my thin little layer of joint compound probably wasn't going to do much to keep the molten metal in place. It was going to spread all over his shop.

He handed me a level. Ha! My amateur eyeballing had been right on. It was perfectly level lengthwise—except for the last foot on one side where it curved like a banana. When making the form, I'd needed to add an extra two feet of wood to the ends of the two-by-fours since they only come in eight-foot lengths. Evidently, one of the additional pieces hadn't been very straight.

Richard was halfway across the shop and unable to see how shoddy it really was. I scrambled around finding rocks to prop under the frame in various spots, but no matter how I tried, I couldn't get it perfect.

"Is it level?" Richard yelled over the furnace, picking up the ladle now filled with forty pounds of molten lead. I imagined the master Roman plumber yelling a similar thing to his apprentice. It was one of those moments you know it's best to tell the truth.

"Yeah," I lied, and then muttered, "Pretty much."

Richard steadily strode over with the leaden ladle, trailing a tail of toxic smoke. A few seconds later, it was over—Richard had poured the ladle's entire contents without spilling a drop and the silvery liquid had spread evenly (seemingly) across the entire length of the frame. Smoke rose all around us and I breathed in the fumes as I bent down to inspect our metalwork.

Bubbles sporadically rose through the lead as it hardened, spurting out tiny puffs of smoke, like miniature volcanoes—Vesuviusinis, if you will.

"That's moisture in the wood coming through. It's going to create severe impurities in your metal. That's why they wouldn't have used wood. A true craftsman wouldn't have stood for this." However, I still had about seven feet of sheet lead that looked usable. I pried the thick sheet of metal out of the frame with a screwdriver. The upper side was a little rough, but the part that had hardened against the plastered wood was smooth and silvery shiny. (Lead starts off sparkly after hardening and only turns dull a few hours later.)

I'd brought a ten-foot section of half-inch PVC pipe to bend the lead sheet around, but now I could see how ludicrous this idea had been. The half-inch-thick sheet looked totally unpliable. All those illustrations of Roman plumbers bending the sheet of lead around a wooden cylinder seemed criminally misleading. I probably wasn't going to be able to bend it a millimeter.

Richard suggested I saw it into smaller, workable sections, which I did back at home. After all was said and done, I was left with two lengths totaling four feet—a far cry from the official ten feet that Roman pipes were supposed to be and the total of twenty feet that I'd hoped for.

Even these short sections were unwilling to bend into pipe shapes, however. Sitting at the end of the driveway to keep the lead away from my house, I tried pliers, a vise, and even Vice-Grips, but it wouldn't budge.

Lisa came out holding Angus.

"Do you think it's safe doing this right here?"

"Of course it is. There's only a few shavings from when I sawed the pieces off."

"But aren't you the same person who vacuumed the entire perimeter of our house to get up the lead paint chips just a few years ago?"

"That was different. Those were lots of little white flakes that would be enticing to the kids." Lead is sweet. (The Romans used it as a sweetener.) Once a kid gets a taste of some sugary lead paint chips, he just can't stop himself.

"Right," she said, and went back inside.

I tried setting the pipe along a piece of wood with half the sheet overhanging and then gently tapped it with a hammer. It started to bend inward, but it also warped the entire length of the pipe-to-be.

My neighbor Helen strolled by with her mom, Martha, who is also a neighbor, walking their shared corgi, Mackey.

"Oh, God, what are you doing now, Hodding?" Helen asked with more than a hint of derision.

"He's making a Roman water pipe, of course," Martha said. She then giggled, just like her six-year-old granddaughter, Willow.

"That's right," I agreed. "Can't you tell?" How on earth did Martha know what I was doing? "I'm having some problems, though."

"Hodding, you've got to set it on the edge and hold it down while you tap part of it," Martha said, demonstrating what I'd been trying to do. "That way you'll keep it from warping."

"I tried that."

"Well, just hit it where it's warping."

They went away. I placed it along the edge and hit it as she suggested. It worked. In about half an hour I had a two-foot long lead pipe. It was about one and a half inches in outside diameter and beautiful: dimpled with hammer marks and the joint barely perceptible. It would have made Rome proud. Of course, the seam still needed to be soldered.

Lisa came back out.

"Oh, wow," she said.

"How do you think it looks?"

"Well, um, it's pretty rough, but it looks . . . good."

"What do you mean by 'rough'?" I asked.

"I gotta go to the bathroom," she said, and hurried inside.

And that's when I knew I'd succeeded. Just the sight of it made her think of running water. A real Roman water pipe, and I had made it, and man, was it heavy. Picking it up with my bare hands, tapping it into the palm of the other, I finally appreciated those oft-repeated stories of someone getting murdered with a lead pipe, like Colonel Mustard in the library in the game Clue. This thing could do some damage.

I went back to Richard's—who, to his credit, did not laugh at my lumpy pipe—so we could solder it shut by pouring molten lead along the lengthwise seam. First I had to build a clay levee around the entire seam so that when Richard poured the molten lead solder, it would stay in place. Once I had done this to Richard's satisfaction—a matter of three tries—he sealed the seam with some molten lead. It worked. A quarter-inch lead joint ran the entire length of the pipe, looking just like some of Professor Hodge's illustrations.

Later, I decided I wanted to test how much pressure it would hold. A typical modern house pipe holds about seventy-five pounds per square inch. A Roman pipe supposedly could hold eighteen psi. I'll never know what mine could hold because it started to fall apart when I tried to jam brass fittings into its ends so I could attach it to a compressor. But just before doing so—for kicks—I pressed one end against an outdoor faucet and turned it on. The water ran strong and undisturbed through the pipe without any leaks.

That had to be a good eighteen psi.

And that's when I started feeling weird—all spacey, disjointed, and achy. But as I said earlier, I don't think it lasted very long.

The Romans themselves, though, didn't fare so well. Many authorities believe that the dementia that many of Rome's later emperors suffered from was caused by generations of lead poisoning. One can't blame the piping, however, because the water rarely sat long enough in the lead pipes to absorb the metal, and even when it did, most of the pipes were lined with calcium deposits from the hard mountain water brought in by the aqueducts. Most of the lead probably entered the Roman citizens' bodies through the lead sweetener they used in their wines, the lead decanters that held the wine, and the lead-lined water tanks that adorned all the homes of the rich and famous. So, while lead may have led to the fall of the Roman Empire, don't blame the plumbers.

Interlude in Bath

What does one do when he's fallen in love but is married with four children? Especially when what he loves is a concept brought to life in the form of metal pipes, drains, and decomposing waste? He crawls around on his hands and knees in Bath while schoolchildren trip over his aging form, snickering at his off behavior.

Thanks to my bout of plumbism, I never got around to figuring out how the Romans joined their pipes *in situ,* but, while still a bit ditzy, I did fly off to England to visit the old Roman bathhouse in Bath so I could see a bona fide Roman lead water pipe. And it turned out not just to be some tiny little pipe like mine, mind you, but an upper-arm-sized one lying alongside the Great Bath—a clear warm pool back in the day when it was covered with a vaulted ceiling but now a great green mucky thing choked with algae that some YMCA pool attendant would love to get his hands on.

The pipe was right there, exposed to every passerby. Some people were even walking on it—in fact, lots of people were walking on it—as they went from the edge of the bath to an arched recess housing some less important ancient relic. I was more than a little shocked. These people were walking over a Roman lead water pipe as if it were some common, modern piece of cast iron.

I lay down next to the pipe. It was slightly recessed in its own shallow culvert and at one time had probably been covered. People stumbled past, some laughing, others oblivious—most of them loud French school kids. When I thought nobody was looking, I ran my hand along the top of the closest section of pipe. I even stuck my finger in a hole and accidentally bent an inner jagged edge that was about to fall off.

I was actually touching the work of a Roman *plumbarius.*

I could hardly believe it, and the pipe I had made really did look like this one. Sort of. It had the same soldered seam and it was lead. Once my euphoria dampened a bit, I could see the differences. These pipe sections hadn't been beaten into shape as I had done but had been bent around a rod. They were very smooth on the outside. And then I noticed why.

They were much, much thinner than mine, even though they were probably six times the diameter. Lesson number 237: don't believe everything you read about Roman lead pipes. They weren't as thick as I'd been led to believe. Thus, they were easier to make. If I had made my sheets of lead as thin as this, even I could have bent them around a rod into the proper shape. At least, it was comforting to think so.

And those joints I had obsessed about? The whole reason I had decided to make the pipes in the first place? They weren't fitted together at all. It looked like the plumbers just set them end to end and then glopped them together with molten lead. I'd read about this technique but had figured it was saved for desperate times when they couldn't do anything fancy, but from the looks of things at Bath, all the joints were done this way. (Cleanly joined pipes have been found at Roman sites throughout Europe, but perhaps they were more the exception than the rule.)

A docent named Diane approached me after I'd been lying in the way for an amount of time that was even too long for the British and asked if she could be of help. Was I all right?

After I assured her that I was more than just all right, Diane told me that the bath's reservoir had been originally lined with lead but that it had been sold to help pay for the restoration of the facility a hundred years ago. "We've got a Roman coffin made of lead in the store. Maybe they'll let you see it if you talk with the curator. The workers who made the pipes probably also made the coffin, right?"

Certainly, I answered, trying to sound like a forthright scholar. I even gave it a little British upper-crust nasality for good measure. A *plumbarius* was a *plumbarius,* whether he was making pipes or lining cisterns or . . . making coffins, although I had no idea if this was really true.

She took me on a short tour of the different rooms, ostensibly to look for more pipes, but I think it was to get me away from the pipe I'd so obviously fallen for. There was a nice stink, thanks to the high sulfur content in the water of the east and west baths, which are enclosed, but we didn't find any more pipes. No matter, though. One pipe was enough for me.

4

How We Do It Now

BEFORE WE LOOK at how we get water in our homes today, let's take a look at how water got here in the first place.

Nearly seventy percent of Earth's surface is water, but it hasn't always been this way. In the beginning, Earth was just a whirling molten sphere spinning around the sun, releasing superheated gases into space, something like an orbiting, overgrown Walter the Farting Dog. It took some one billion years before anything else of significance occurred. Over time, Earth did cool, and as it did, the gases being released started to form an atmosphere because, being cooler, they didn't rise as high. Some of those gases were water vapors and they bound together high above the ground as clouds. As these clouds formed, Earth spun happily along, unaware that its surface was about to receive an onslaught that would begin an alteration of its surface and structure that continues to this very moment.

When the clouds cooled, their vapors solidified into rain, which fell down to the earth. This happened again and again, for eons, and Earth's surface became inundated with water. The clouds and the rain caused Earth to cool even more and a crusty surface formed.

The same water that was here then is still here today—no matter how much we use and abuse it. The amount of water we have hasn't really changed—just its salinity and quality.

The water vapors that rose into the atmosphere had some salts in them but not enough to make the fledgling oceans as salty as they are now. In fact, the primordial oceans of a few billion years ago were nowhere near as salty as today's. Most of us would have been able to drink a glass from the oceans without pause. They were a vast, sugarless Gatorade.

If the oceans had stayed salt-free, water supply wouldn't be much of an issue today, but, of course, they changed dramatically over time. Most scientists believe that as rain fell, it eroded the rocks that had formed on Earth's surface, dissolving their minerals into the growing seas. A large portion of these minerals was some form of sodium. Over eons, the oceans grew saltier because more salts stayed a part of the liquid mix than settled to the bottom. Even today, although the amount of dissolved salts being added has slowed down a bit, rivers and streams dump close to four billion tons of salts into the oceans annually. The oceans also continue to get saltier through evaporation. When the sun heats ocean water, the water molecules vaporize and the salts mainly stay behind. It would take much higher temperatures to vaporize the salts as well.

As the cloud vapors cool, they turn into solids—rain—and fall from the sky. The fallen water accumulates—either in oceans or lakes—and the whole thing starts over. This pattern is called the hydrologic cycle. Some scientists, however, believe that thousands of house-sized, water-laden comets bombard our atmosphere daily and are thus actually constantly increasing our water supply. Most other scientists think these comets contain a kind of water significantly dissimilar to what we call water—sort of the way a Hydrox isn't really the same thing as an Oreo.

So, we have all this water. But how does this ancient element (one of the Big Four: Water, Fire, Air, and

Earth) get into our homes? As mentioned, rain or snow falls and collects in lakes, natural reservoirs, and aquifers. Tapping either underground sources or aboveground reservoirs (currently more than half of the American population gets its water from underground sources), we typically pipe it via large twenty- to thirty-inch-diameter pipes to a treatment plant. These pipes are usually made of ductile iron or steel and are lined with cement, although some municipalities have replaced them with heavy-duty PVC because the plastic is thought to be more resistant to the corrosive elements found in the ground. These lines are often not pressurized, since the treatment plant is usually located at a lower elevation than the reservoir and it doesn't matter how fast the water arrives—just that it does. Up to this point, things haven't changed much since the Roman days. In fact, in Massachusetts and some other states, these initial water delivery systems are still called aqueducts. The controversial Los Angeles water system—ever see *Chinatown*?—is called an aqueduct, delivering water from hundreds of miles away.

What happens next, though, is a huge departure from the past. The water is filtered and then chemically restructured beyond recognition. It's still water, of course, but nothing like it was. (In some states, if the water source is essentially free of harmful bacteria and other contaminants and will remain so in the foreseeable future, the water supplier does not have to filter the water.) There are dozens of ways to run water through a treatment facility, but essentially what happens in all of them is that the water goes into a humongous tank and gets zapped with heavy doses of chlorine—levels way past what the EPA says is safe for drinking—to kill anything that might be alive. However, while the water travels through the different chambers of the treatment tank, the excess chlorine gets consumed by the bacteria-killing process and drops down to one-half to one part per million, the nationally recommended level. As the water flows out of the tank, it is often dosed with fluoride for our teeth and a corrosion inhibitor. The EPA has something called the lead and copper rule that requires water utilities to monitor the amount of heavy metals in their customers' homes. Aggressive water, as the utility guys call it, can exacerbate the leaching of metals into households. Water that is too acidic or has too much chlorine is aggressive and breaks down metal. If you have lead solder in your pipes, as most older homes do, it can deliver that lead directly to your body. It also breaks down copper, ingesting too much of which can lead to kidney and liver damage. Thus, we have the EPA-mandated corrosion inhibitors. The water is now almost ready to be delivered to you, the consumer, except the fluoride and corrosion inhibitor treatments have depressed the pH level, so caustic soda is added to bring the water back to neutral—also federally mandated.

The upside to chemical treatment is that it's usually cheap. The downside is that when water quality isn't good, it requires more chemicals and more monitoring, making it not so cheap and not so safe. Many people have an intolerance to chlorine at any level. Also, water that hasn't been treated chemically simply tastes a hell of a lot better. That's why Americans currently spend more than $8 billion a year on bottled water.

Anyway, once treated, water travels by mains throughout your town or city. Most mains laid from the mid-1800s to the 1950s were cast iron, but they are generally being replaced with ductile iron lined with cement. Cast iron gets brittle over time while ductile iron will theoretically stay resilient for a much longer period.* In the 1970s, some municipalities replaced their old iron mains with PVC because it was cheaper than iron, it is easier to maintain pressure in PVC thanks to less friction, and because PVC resists corrosion better than iron. As manufacturing iron became

*_Ductile_ means both malleable and tough. Ductile iron has twice the strength of cast iron, mostly brought about by the introduction of magnesium to molten iron during its manufacture. Ductile iron, introduced into the marketplace in 1955, is the pipe of choice.

more affordable, however, most areas went back to iron primarily because PVC leaches contaminants through the pipe wall. For example, if PVC is laid through soil that is soaked in gasoline, like at an old filling station, the gasoline will go through the PVC and into the water system

The size of a water main depends on the size of the city being served, how many houses and businesses there are, the distance and elevation the water has to travel, and maybe even the power of the water supplier's pump. The mains leaving my local treatment plant are either twenty-inch or sixteen-inch in internal diameter. New York City's largest mains, by contrast, are eighty-four inches. The three towns served by my water provider go through 3 million gallons of water a day. New Yorkers use 1.5 billion.

As the main travels under and alongside local streets, it is commonly reduced to either six- or eight-inch pipes—depending on the number of houses and businesses on a street and their expected water consumption. This is fairly standard neighborhood sizing—found both in my town and New York. Each pipe is usually twenty feet long and, in the case of the six-inch iron pipe, weighs about five hundred pounds.

Shutoffs are placed every thousand feet so leaks can be fixed. This figure is merely an ideal, however. Old mains might go up to a mile without a shutoff of any kind, making life pretty difficult when things go wrong. Go on the Internet and type in "leaks main water lines" and you'll see even the smallest cities have hundreds of leaks each year. Why? Our Earth is in a constant state of flux, and it has the regenerative habit of composting anything buried in it. A mile-long main is about as secure and stable as a first marriage. There are 150 miles of mains in my water region serving eight thousand homes and businesses. My utility has to fix about ten breaks and twenty-five leaks every year. New York has 6,000 miles of mains, serving nine million people, and has thousands of leaks and breaks that must be stopped annually.

Houses and businesses connect to the main with a pipe sized to keep the pressure at least forty psi but no greater than eighty psi. Anything beyond eighty psi will blow out fixtures and valves. The ideal pressure is somewhere around fifty psi.

Before that pressurized water gets into your house, though, let's take a quick look at who is bringing it to you. You might assume that it's your city or some sort of state or county authority doing the providing—that it's a public utility—and if you live in the U.S. or Canada, this is generally correct. As of the early 1990s, public utilities served eight-five percent of the American population and virtually all of Canada's. In other parts of the world, from France to Argentina, though, private companies supply most of the water. Privatization or corporatizing of water supply is increasing worldwide because of poor infrastructure, shortage of government funds, and general mismanagement. In 1989, England turned over its entire system—both supply and sewage treatment—to private companies, the largest turnover of government water management to private hands in the world, affecting some fifty-five million people. Only time will tell if this system is best for England as a whole. And, given time, it's likely many of these privately held companies will go back to public control. While early water supply, as in Rome, was provided by the state, ever since it's been more a game of hot potato, as opposed to tug-of-war, between public utilities and private companies: each thinking they want it, only to realize decades later that they can't handle it. This is also true for sewage and waste water treatment. London alone has gone from private to public to private in the last hundred years for both supply and treatment.

In the U.S., there appears to be a current trend to turn over water supply to private companies as well, and for those of us who think of water as something of a guaranteed right of all citizens, this is fundamentally unsettling. It's not that the private companies are abusing their power and overcharging us: public utilities commissions still control the price the companies can charge. Rather, it's because next to air, water is our most basic need. You can only live a week, at most, without it and you need to consume two and a half quarts daily through liquid or food to remain healthy. (It's a modern myth that you need a gallon of drinking water a day, by the way.) Understandably, people worry about others playing God with it.

Guess who was getting into the water business just before everything went down the drain? Enron. In the late nineties, Enron was buying up water rights and water companies to start a water-trading market, just as it'd done with natural gas and electricity. Water to the highest bidder!

World water consumption will triple over the next fifty years and scientists and sociologists have begun warning about future water wars. As the world's population grows, the ability to supply enough potable water is going to be next to impossible. Current overexploitation of certain water reservoirs is already threatening to result in violence, specifically in North Africa, where Libya has recently begun tapping into the Nubian aquifer, which lies beneath Libya, Egypt, Chad, and the Sudan. International mediation has calmed things down for now, but imagine that same scenario with Enron-esque companies in charge—all over the world.

Okay, back to supply. The water is just outside your yard, waiting to be used. According to the American Water Works Association, on average each American uses 69 gallons a day—11½ on the shower, 15 on the washing machine, 1 on the dishwasher, 18½ on the toilet (the biggest consumer of water), 1 on a bath (guess we don't bathe that often), 9½ on leaks, about 11 on the sink, and another 1½ on unnamed things. So, how do you get the privilege of using 9½ gallons on leaks on any given day?

Your utility attaches a three-quarter-inch copper pipe to the main and runs it a few feet onto your property to something they call a "curb stop." This is the main valve that regulates the supply to your house and it has a shutoff that generally only the utility can operate. From the curb stop on, everything is your plumber's problem.

Your plumber attaches a three-quarter-inch pipe to the curb stop and runs it underground—five feet on account of frost in my area, two feet or so in warmer areas—to your house. The pipe goes through the foundation and at this point you should find a shutoff valve so you can cut off the water to your house when doing repairs. This is usually a ball valve made of brass, it being more durable than steel or copper. It's called a ball valve because it contains a stainless-steel ball, the size of the opening, to block the flow. Turning the handle turns the ball, exposing a side of the ball that has a tunnel going through it. Water can now go through.

Before the water branches off to your water heater or anywhere else, it encounters a few different gadgets that you may not recognize. The first of these is the water meter; it's easy to recognize, with its ever-increasing numbers. Next, in places where the pressure in the main is too great, you might find a pressure-reducing valve—a simple device that, using a spring, reduces the flow and thus the pressure in your system. After that, you have a backflow preventer. This device keeps nonpotable water from entering the town's water system when something's gone wrong with the water main. Say there's a fire next door. The pump truck comes and starts hosing down the house. It uses so much water, you lose pressure in your lines. If you happen to have a backed up toilet that is so backed-up that the backed-up water is somehow touching the toilet's water inlet, all the stuff in your toilet could conceivably get sucked down into the main, filling the void, so to speak. When the main is working again, your toilet water and even what was in the toilet's trap, will become part of the "fresh" water supply. Since you don't want to drink sewer water, nor do your neighbors, most municipalities require that you have a backflow preventer.

Now, under an ideal fifty psi of pressure, the water separates into two lines. One becomes the main cold line. The other travels to your hot water heater. Then the hot water meets up again with your cold water supply line at a manifold and from there, separate cold and hot supply lines connect to your various fixtures—as toilets, lavatories, baths, and refrigerators are known in the trade.

This all seems simple enough, but pretend you want to connect a pressure-reducing valve and an expansion tank to your water supply line because excess pressure has been making the relief valves on your boiler spurt out half a gallon of water a day onto your furnace room floor for the past three years. You have excess pressure in your system due to no fault of your own but because a rich fatcat lives down the street. His house is so big and has so many bathrooms that the main going down your road is bigger than all the other mains on streets similar to yours. (This is, of course, just a hypothetical situation.) Say you discovered the leaking boiler more than a year ago and have since been collecting the water in a rectangular piece of Tupperware and emptying it into a red five-gallon bucket on a daily basis.

Let's also say you sometimes forget to do this and so the water on the furnace room floor often runs to the lowest point, which happens to be where your new shower drain and waste pipe go down to the basement. (Again, this is a hypothetical situation and bears no resemblance to anything in anyone's real life.) The water drips down through the plywood subfloor surrounding the waste pipe and if you don't do something about it, pretty soon the $30,000 renovation of the mudroom/furnace room/bathroom wing that you're still paying for is going to be ruined. Let's also say that your wife keeps suggesting you get George, um, I mean, a plumber, over to fix it but every time you bring it up with the plumber, he suggests you do it yourself so you can really know what it's like to be a plumber. And then, finally, let's say one day you decide to do it, unable to bear the shame any longer.

You go get twelve feet of Type L three-quarter-inch copper tubing. You buy type L because it's more expensive than Type M and you figure it's got to be better. In fact, there are three types of copper tubing: K, L, and M. K is the thickest and should be used from the curb stop to your house. M is the thinnest and is what is usually used throughout your home. L falls somewhere between the two and can be used for a variety of purposes. You buy a wide assortment of copper adapters that go from three-quarter-inch to half-inch with threaded and nonthreaded openings. (The backflow preventer has a threaded fitting.) You buy a new blowtorch because the last one you owned failed you miserably when you tried to fix the plumbing at your uncle's summer shack, an effort that ended with a call to your plumber friend to rescue you. You also buy a thick roll of lead-free solder and a small bit of flux; flux is a chemical compound that allows metals to join together by excluding oxygen from the solder site. The presence of oxygen would lead to oxidation, which deters metallurgical matrimony. (The Romans used olive oil for this.) You buy a hand tool that has wire brushes for smoothing and cleaning out the insides of the pipes and fittings and also has a device that cleans the outside of the pipes. You just push the tool into the fitting, or over the pipe end, twist the tool, and it scours the surface with steel bristles. You get Teflon tape to seal the threaded joints.

You rush back home because you've already cut off the water supply to the house and even cut the pipe ahead of time—an interesting decision considering your wife is currently at home with your twenty-month-old son. He's your fourth child and you'd think you would have learned by now but you haven't. She gives you "the look" from the kitchen window as you lower yourself into the basement. These things used to be cute, and even after eight years of mostly sleepless nights she still manages to be amused on many occasions. This just happens not to be one of them. You find yourself wondering if this decision to "sweat"—plumber's term for soldering pipes together—your own pipes is going to push things over the edge.

You twist and squeeze yourself into the basement that no plumber likes to enter because it's more of a crawl space than anything else. You scoff at their precious behavior and trot, bent forward with your torso at a right angle to your legs, to the rear of the basement where the supply line is located. On the way, you bash your spine on a hunk of metal protruding below the floor joists. You scream and immediately stop scoffing at the precious plumbers.

Finally, it's time to begin. Having learned a thing or two from your plumber friends, you measure everything twice. You then cut each piece, marking the cut line with a permanent marker. Actually, you've forgotten the marker up on the kitchen table. Not wanting to worry your wife with your twelfth return from the basement (you've actually had to go to the hardware store four times already for forgotten items), you cleverly use the edge of a dull screwdriver to mark each cut. Then you fit everything together without soldering. Again, something you've picked up from the guys.

After that, it's just a matter of taking it back apart and sweating each joint together. You're not too sure why it's called "sweating," but you love using the terminology: "Hey, baby, I sweated sixteen joints, an expansion tank, and a ratcheted striptip transmogrifier today. How about you?" Later, you ask a plumber why it's called sweating and he says it's because the blowtorch makes you sweat so much. Perhaps, but maybe it's because heated metal turns the metal solder liquid and draws it up just like a heated body draws out sweat. Be that as it may, the joint that is going to connect one piece to another gets heated up until solder melts when applied to the heated area. You then heat the other side of the joint until solder melts there, too, to make sure it's hot enough everywhere. Next, you heat the whole thing for another second or two, being careful not to burn the piece of wood behind the pipe again; you burned it a little already but haven't told anybody about it. Now the joint is so hot that when you apply the solder to the edge of the fitting, it gets sucked underneath—a capillary effect, it's called—and seals the space between the fitting and the pipe. Then you quickly wipe a wet cloth around the hot solder, which makes it smooth. The smoother the seated joint, the more points you get in the eyes of your fellow plumbers. And it's all about those points at this stage, isn't it, even though you're not really a plumber?

You do this for fourteen joints. You have to make a ninety-degree detour in the supply line to fit in the expansion tank, building what looks like an upside-down, squared-off U in the line. Two hours later, you cut the water back on and, miracle upon miracle, it doesn't leak anywhere. You're the man.

You start packing up your leather-trimmed cloth tool bag that no real plumber would ever be caught using when out of the corner of your eye you notice a leak, but thank God it's not where you've been working. The main shutoff valve has corroded and water is drip, drip, dripping off the handle. It's not your fault, so you turn to leave. Then you remember it's your house.

You call your water supplier and the next morning they come over and shut off your water. You tell them the plumber will be at your house shortly and could they please return to cut on the water in two hours or so. The nice man who's been working with the water company for forty years says, "No problem." He drives off in his van and you run down into the basement, hitting your back again on the hunk of metal. As you did the day before, you carefully measure, mark, cut, and solder. In less than half an hour, you're done. You go upstairs, brag to your wife, squeeze your baby boy's rosy cheeks, and feel deservedly cocky.

The water guy comes back on time, cuts on the water, and everything goes perfectly. There's nothing to this sweating stuff. Any fool can do it.

Oh, except that there's a huge hole on the backside of one of the soldered joints for the shutoff valve and water is spraying twelve feet across the basement and up into the insulation in the floor joists above. You go running out to the water guy and tell him the plumbers have somehow messed up. There's a small leak. He shuts off the water. He says he'll be back in twenty minutes and to get the plumbers back on the job. That way, if there's a problem again, they can fix it while he's there. He looks at you sort of funny, like he might know who the plumber really is. Or is that just your guilty conscience?

He leaves. You dash down to the basement, hitting your back again, open up the closest water valve to drain the line, and then desperately try to sweat this tiny one-millimeter-long gap closed, but water keeps bubbling through the crack, cooling off the joint just enough so that the solder doesn't take. You try it again and again. Tears well up. And then you hear the dog barking.

The water guy has returned. He comes into the basement and sees you at the pipes. You tell him you couldn't reach the plumbers. He grunts and works his way over to you. For a while he lets you attempt to fix the situation, but after a couple of minutes he gently grabs the torch from you and applies the flame to the joint, much closer than you've ever done. A few seconds pass and he motions for you to push the solder against the gap. Instantly, it sucks in the solder, closing the space for good.

He smiles and says it just wasn't hot enough. When he cuts the water on this time, there's not a leak in the house.

Traditionally, for more than a hundred years now, that's how your pipes have been connected together—albeit more professionally. Recently, however, a lot of plumbers have switched from copper pipes to cross-linked polyethylene (PEX) pipes. Unlike PVC plastic, PEX is considered safe for supplying water, as it has not been shown to be carcinogenic, and it is much simpler and faster to lay out than copper. You just run a hose from Point A to Point B, securing it to a joist or stud every twelve inches or so, clamp it to an adapter at the fixture, and you're done.

So, whether it's through plastic or copper, that's how the water gets from the reservoir to your sink, waiting at full pressure at your faucet. The reason it doesn't come pouring out like in Roman days is because your faucet has a ball valve in it, just like the shutoff valve in your basement. When you turn the handle of the faucet, it turns the ball and exposes the part with the hole running through it to the pressurized water. The water then races through the hole and out of your faucet into your sink. You wash up and then turn the handle off. The ball rotates and the water stops.

From there, the water goes down your sink drain and enters a water trap. Every single fixture in your house—from your dishwasher to your toilet—must connect to a water trap. In the past, these traps were S-shaped and thus were cleverly called S-traps. In ideal circumstances, the S-traps did a splendid job of creating the necessary water seal that would block sewer gases from entering your home, but in certain situations atmospheric pressure could cause water to be siphoned out of the trap, allowing deadly sewer gas in. Today's P-traps account for this effect and hold water no matter what the changes are in atmospheric pressure.

After the trap, the drainage line runs at a slight downward slope—a quarter-inch drop for every foot traveled—into your wall, where it makes a ninety-degree turn at a T and drops straight down to your basement. The upward part of the T connects to the vent pipe, which runs up to your attic, where it meets up with all the other vent pipes and connects to the vent stack—that black pipe jutting out of your roof. Every single plumbing fixture must have its own vent pipe. There are all sorts of rules as to where the vent stack can be placed in relation to second-floor doors and windows, but suffice it to say that it must be far away because of the deadly gases it releases.

Returning to the wastewater going down to your basement, just for the fun of it, let's say that you're not washing your hands but flushing your toilet. You pull the handle. Typically, this handle is attached to an eight-inch plastic or metal arm called a trip lever that dangles a short chain from its end. The chain is attached to a plug, called a stopple or a tank ball. When you push down on the flush handle, the metal arm lifts the stopple from a two-inch-wide drain. The stopple stays up until nearly all of the water has left the holding tank. While water rushes into your bowl via gravity, that roundish thing in your tank called a float ball drops to the bottom and triggers the cold water supply line to turn on and refill your toilet tank via the filler tube. As the water refills, the float rises to the point your plumber has determined, usually per manufacturer's instructions, that is high enough to make the toilet function properly. Then it shuts off the supply line.

An overfill pipe sits above the waterline in case something goes wrong and allows any water that goes higher than the fill line to drain out.

Meanwhile, the toilet bowl—that's the part you sit on—has a built-in water trap that feeds into a PVC drain pipe, called a waste pipe. Water comes flooding into your bowl and as it does it pushes water over the lip, or dam, of the trap. As more and more water flows over this dam, all of the air that's been in the trap is pushed out as well, creating a vacuum, momentarily draining the bowl. Then the trap partially fills back up with water. Simultaneously, as the discharge flows through your PVC pipes, the toilet bowl fills up enough to catch your next deposit and provide a wall of water that keeps sewer gases from rising up through the bowl.

The wastewater, whether it's from your toilet or anywhere else, travels through PVC pipes that start off at a two-and-a-half-inch diameter and increase to four-inch when they reach the basement or crawl space. This pipe must also run at a quarter-inch pitch as it travels through the bottom of your house and continue at this pitch underground until it meets up with the main sewer line, which more often than not runs adjacent to the main supply line. These days sewer lines are increasingly made of plastic, but older pipes were made of anything from lead to iron to cement to brick. Since this system is not pressurized, more flexibility was possible when choosing material for the sewer line.

At this point this is an entirely gravity-controlled system, so the pipe going down your street and toward the town sewage treatment plant must continue downhill if your waste is going to make it all the way to the treatment plant. Of course, sometimes this is impossible. In those cases, sewer lines are augmented by pumping stations that draw the sewage up a few feet and then release it into pipes placed higher underground. The sewage then flows to the optimistically named "water works" plant, where it is treated.

The Clean Water Act of 1972 and the Safe Drinking Water Act of 1974 require secondary and tertiary treatment for all American sewage. Before these acts, most municipalities only used primary treatment, which is simply the separation of solids and liquids. In many cases, solids were sloughed off to sewer farms, while liquids were dumped into the nearest waterway. Secondary and tertiary treatment actually makes the water almost potable by killing all harmful bacteria and returning the water to a pH-neutral level through the use of chemicals. Many cities, while not offering it as a beverage, provide tertiary-treated wastewater for landscaping purposes. Goleta Valley, California, population seventy-seven thousand, saves 300 million gallons of water a year this way, while the East Bay District in San Francisco recycles some 5.5 billion gallons.

This cleaned-up wastewater, whether it's used directly to water your lawn or, as more commonly happens, is dumped into a local river or bay, over time reenters the atmosphere by either evaporation or evapotranspiration, the process by which water evaporates from the soil and transpires through plants into the atmosphere. Then it cools off in the cloud layer and returns to the ground as rain, sleet, snow, or hail for us to use all over again.

And that's how it works.

5

FOUL

"FAIR AND FOUL are near of kin / And fair needs foul," cries Yeats's Crazy Jane in "Crazy Jane Talks with the Bishop."

"Foul needs fair to become fair again," I respond, thinking of sewer water treatment.

"But Love has pitched his mansion in / The place of excrement / For nothing can be sole or whole / That has not been rent."

Enough, I say.

We know from the architectural remains of the Harappan civilization that waste removal has been a concern for quite some time. The unparalleled Harappan system of brick gutters that took waste from the home and dropped it in the nearest river established a path for future societies to duplicate. If you're going to live together, get the waste out of town. It was a simple, straightforward pathway. Yet it was hardly followed. Most subsequent civilizations did not bother with such practical amenities, except for the Romans, and by the time civilization in the form of Christianity had drifted toward Western Europe, waste removal had fallen upon very dark times indeed. The streets of Europe were, quite literally, cesspits.

Although people gravitated toward urban living during the early Middle Ages, they did not always maintain the Roman-era sewers or feel it necessary to improve upon them. In fact, people had nowhere to relieve themselves, except in pots kept hidden in corners and in some public latrines that were built directly over rivers or cesspits.

(The anomaly during these backward times were the Welsh. They had learned a thing or two from the Romans—not so much in the area of plumbing but in the uses of human waste. In the tenth century, Hywell the Good created the post of Minister of Urine under the notion that urine was too precious a commodity to continue wasting. Welsh communities also appointed *bismaers,* or dung mayors, who were in charge of buying and selling dung for the town. History books do not tell us how the Minister of Urine collected any of the precious substance, but perhaps the Welsh retained the Roman practice of placing urine pots at various street corners.)

Over time, as the gentry multiplied, statelier homes demanded something more than just a hidden pot. Across Europe, by the middle of the Middle Ages, a new room called a garderobe, originally a French word meaning a room to keep clothes in, could be found in all the more modern castles and homes. The garderobes were separate little rooms that did not store clothes but instead housed a seat of stone or wood with a hole in the middle. The contents of the garderobes often dropped straight onto the road below or into moats—so, please, vanquish all your leftover romantic images of medieval castles from childhood fairy tales or Disney movies.

Didn't you always wonder what the big deal was concerning moats? Any dope could just swim or wade across them, right? Wrong. Moats were simply circular cesspits and were in actuality some of the earliest forms of germ warfare, practically bubbling with *E. coli,* cholera, and other such lovelies. If you fell in one while storming a castle, there was more than a good chance it would be your last raid.

Up to and past the days of King Henry VIII, garderobes reigned supreme. Henry's courtiers used the Great House of Easement at Hampton Court. It was set at the edge of the palace beside the Thames: two rows of oak seats on two separate floors with holes cut in the wood at two-foot intervals. Chutes made of brick or stone carried the waste down to a brick basin, which then allowed the liquid effluent to go directly into the river. The gongfermors, men whose job it was to collect the sewage, took care of the rest. Since there was no water flushing down the waste and the chutes were connected directly to the building, it must have been quite stinky.

What did those without garderobes do? As time went on, collecting waste in a pot of some sort, the ever-increasingly ubiquitous chamber pot, that is, became a harsh reality of urban living. After a day's worth of defecation, the happy homeowner or tenant simply tossed the contents out the window, usually at night, making the world's oldest profession even more dangerous.

In France, the tossers took pity on their fellow humans and when unloading their pots, they would call out, "*Garde l'eau!*"—meaning, "Look out for the water!" Over time the streets became choked with excrement. The practice grew so horrific that by 1539,

François I, the King of France, proclaimed by decree that he "makes known to all present and all to come our displeasure at the considerable deterioration visited upon our good city of Paris and its surroundings, which has in a great many places so degenerated into ruin and destruction that one cannot journey through it either by carriage or on horseback without meeting with great peril and inconvenience." He went on to outlaw all tossing of waste, but to no avail. City streets remained deadly dangerous for centuries to come.

A bastardized version of *"Garde l'eau"* crossed the channel and throughout the British Isles you would hear people yelling, "Gardy loo!" just before the torrent was unleashed. Some have suggested that this is where the British slang for a water closet, a loo, derives.

In Tobias Smollett's *The Expedition of Humphrey Clinker,* a humorous epistolary novel about traveling in England published in 1771, one of the protagonists, Winifred Jenkins, provides a lasting image of such a situation. "There is nothing for poor sarvants, but a barrel with a pair of tongs thrown a-cross; and all the chairs in the family are emptied into this here barrel once a-day; and at ten o'clock at night the whole cargo is flung out of a back windore that looks into some street or lane, and the maid calls *gardy loo* to the passengers, which signifies *Lord have mercy upon you!*

and this is done every night in every house in Haddingborrough; so you may guess, Mary Jones, what a sweet savour comes from such a number of profuming pans; but they say it is wholesome, and, truly, I believe it is."

Smollett's title character, Humphrey Clinker, confirms this situation while shedding a little light on water supply: "The water is brought in leaden pipes from a mountain in the neighbourhood, to a cistern on the Castle hill, from whence it is distributed to public conduits in different parts of the city. From these it is carried in barrels, on the backs of male and female porters, up two, three, four, five, six, seven, and eight pair of stairs, for the use of particular families. Every story is a complete house, occupied by a separate family; and the stair being common to them all, is generally left in a very filthy condition; a man must tread with great circumspection to get safe housed with unpolluted shoes. Nothing can form a stronger contrast, than the difference betwixt the outside and inside of the door; for the good women of this metropolis are remarkably nice in the ornaments and propriety of their apartments, as if they were resolved to transfer the imputation from the individual to the public. You are no stranger to their method of discharging all their impurities from their windows, at a certain hour of the night, as the custom is in Spain, Portugal, and some parts of France and Italy—a practice to which I can by no means be reconciled; for notwithstanding all the care that is taken by their scavengers to remove this nuisance every morning by break of day, enough still remains to offend the eyes, as well as other organs of those whom use has not hardened against all delicacy of sensation."

Walking around the streets of England, France, and most other European countries remained a distasteful business for some time, although it did give rise to the profession of being a nightman or scavenger, latter terms for a gongfermor. These brave, well-paid souls carted off human waste at night, and the material they removed became known as night soil. The manure was then sold to farmers in the nearby countryside. Present-day travelers to China might remark in an amused tone that some enterprising Chinese pay money for their fellow citizens' waste and then turn around and sell it for a higher price to farmers to use as fertilizer. They say this as if such a thing is not only gross and preposterous but utterly foreign to our Western ways, which, of course, is far from the truth. Night soil has been a commercial product for countless centuries and was still traded in the U.S. well into the 1800s.

(We still have nightmen today, of course, but instead of lurking through the streets at night, they empty septic tanks with pumps at any time of day. Tellingly, I believe, our nightmen do not sell the manure to farmers but instead simply dump it on isolated land or unload it at municipal treatment plants.)

Municipal ordinances throughout Europe, similar to that of France's King François, had been enacted numerous times in the centuries leading up to his 1539 decree, some of which required city dwellers to build cesspits for holding their waste until it could be removed. Buckets of waste were dumped into the pits by the city dwellers and then as ordained by law, gongfermors would remove the waste with long-handled scoops.

In fact, the earliest building code ever enacted in England concerned the building of cesspools. London's first mayor, Henry Fitzailwyn, decreed in 1189 that a cesspool, euphemistically called the "necessary chamber" by Fitzailwyn, "should be 2½ feet from the neighboring building if it was made of stone and at least 3½ feet if made of other materials." The depth of the cesspit depended on the size of the family, but the size did not have to be gargantuan since, absent water closets, no extra water was entering the pit. Usually the pits were lined with some kind of stone, which not only forestalled the collapse of the pit but also slowed down the seepage of liquids into the surrounding soil, although such seepage was part of the plan. Cesspools, when properly maintained, had certain advantages over dumping raw sewage straight into the Thames or any other river, in that they allowed the town's water source to remain clean. However, they were not always regularly emptied, since nightmen asked a fairly handsome sum. In cases where someone's pit was built too close to another's home, the results were disgusting. A certain William Sprot was recorded complaining to the Assizes, a periodic court session, in 1328 that his neighbors' "cloaca" had spilled over his wall and into his yard.

Meanwhile, plumbing was advancing, albeit very slowly. Plumbers made and installed lead pipes, mostly in monasteries but also in some castles. They made these pipes just as the Romans had, from sheets of lead folded and beaten into shape around a wooden tube. At times, though, they created miles-long pipelines. The first recorded piped water supply to London occurred in 1237. A Gilbert de Sandford granted the city the use of the water on his land at Tyburn (in present-day Marylebone) some three and a half miles away, as well as the right to install and repair the necessary lead pipes.

More often than not, the pipes were simple affairs, supplying a sink from a private cistern or a much larger one provided by the town. These large cisterns were fed by lead or more frequently wood pipelines and quite often the cisterns would be lined with lead. Such a cistern at Cheapside during this period fed local houses, and users paid five to six shillings a year for the privilege.

Over time, though, the plumber's work grew to be in such great demand that the first plumber's order was formed in England in 1612 under King James I, arguably the most enlightened monarch who ever reigned over all of England (and yes, I'm suggesting this merely because he oversaw the formation of a plumber's society). The Worshipful Company of Plumbers was incorporated by charter and boasted itself of "a large and very memorable antiquity, remaining a fellowship or brotherhood by the name of Plumbers." The charter also made it unlawful for anybody to "use the art and mystery" of plumbing without being a member of the brotherhood.

As London grew, it needed more water, and in 1613 a Sir Hugh Myddleton built a thirty-eight-mile artificial channel, called the New River, to supply the city's homes with water from Hertfordshire. It took six hundred men five years to build the open channel, which served various parts of the city with wooden and lead mains, and the cost of a year's supply of water for a typical household went up to twenty-six shillings—four times the cost of water from the old Cheapside cistern.

While the New was a momentous achievement, it was merely an open channel. Over on the Continent, they had already begun to make and use cast-iron pipes. Apparently the first-ever cast iron was made in Ghent in 1313 for the production of cannons. Some time afterward, water suppliers used this technology for making pipes, and in 1445 cast-iron pipes supplied Dillenburg Castle in Germany. By the mid-1600s, the French were using cast-iron pipes as well, and a five-mile-long cast-iron pipe supplied Versailles with fresh water. The first cast-iron pipe wasn't laid in London until the mid-1700s. Wooden mains predominated in England as the preferred water carriers because wood was so much cheaper, not because it was a better material.

Slowly cast iron did take over as production methods improved and iron pipe became cheaper to make. Also, unlike wood and lead, it could withstand great pressures and did not corrode underground. (That's also why it is still the most popular conduit today.) Similar advances occurred in lead pipe production, and since lead was much more flexible, it was the metal of choice in urban situations.

By the early 1800s, plumbing in most houses consisted of a cast-iron or a lead pipe that was connected to a town's or neighborhood's water main. The pipe went either directly to a sink in or adjacent to the kitchen, or into a cistern made of bricks lined with plaster, that was often located in the basement; water was then pumped via lead pipes from the cistern to the kitchen sink. This end was fitted with a brass valve in some cases, but in many it had no stop cock. The town or city only provided water two or three times a week and for only a few hours. You simply used the water while it was there.

As supply improved, people put in more sinks and eventually water closets. Water supply had become a consistent, around-the-clock affair.

But what a tangled mess it was. Houses had lead waste pipes running all over and at first these lines merely dumped into a cesspool or, infrequently, a sewer line. As explosions occurred and the stench of rotting waste filled people's homes, they realized they needed some sort of water trap to seal out the smell and gas. Eventually, manufacturers started producing five-inch-tall lead traps shaped like little barrels. Plumbers would then bring all the lines from all over the house and have them pass through this trap, creating these confusing-to-everybody-except-the-plumber sexopuses and octopuses. People didn't care, however. They were getting their water and the waste was leaving the house.

But was it going far enough away?

6

A Tour of the London Sewer

If things seemed scary during the Middle Ages, with waste flying out of windows at all hours, the early 1800s were an outright nightmare.

As London grew, so did the need to cover the city's rivers and streams for the simple fact that more surface area was needed. Cover the streams, ditches, and man-made canals and you have more places to erect buildings. Once concealed, though, these waterways quickly became flowing cesspools. Not only did the real cesspools overrun and spill into them, but as the night-men's wages jumped to six pennies a night, people found it much easier to unload their waste straight into the covered waterways. Since they were covered, who cared, and more important, who was going to see it? By 1815, the practice had become so common that the city had no choice except to allow people to connect household drainage to the sewers. Interestingly, these waterways had been called sewers for hundreds of years—meaning waterways maintained for drainage—but with the advent of becoming covered, *sewer* took on its modern meaning. All of these new sewers dumped straight into the Thames.

Now, while keeping that situation in mind, realize that the sewers did not rid the city of cesspools. It was still easier to make a family cesspit, and by the early 1800s, more than two hundred thousand dotted the city's alleyways and yards. As emptying them became more and more expensive, they got fuller and fuller until they were either abandoned or spilled into yards and basements. Then the waste simply sat there, spreading stench and, unbeknownst to most people, disease throughout the city.

Concurrently, water closets, having become increasingly popular, emptied more and more putrid water into the overfilling cesspits or straight into the sewers and were blamed for more horrors than they solved. "Almost coincidentally with the appearance of epidemic cholera," a government statistician wrote of an

1849 outbreak, "and with the striking increase of diarrhea in England, was the introduction into general use of the water closet system, which had the advantage of carrying nightsoil out of the house but the incidental and not necessary disadvantage of discharging it into the rivers from which the supply was drawn."

England's infant mortality rate rose close to fifty percent. Babies were dying of infected drinking water, perhaps killed by their own parents' waste, since, quite often, drinking supplies taken from the Thames were a stone's throw from a sewage discharge. Fishing, once a thriving industry along the Thames, became a practice of only the desperate. Not a single salmon had been caught for more than three decades and the only thing living in the Thames, besides bacteria, were eels. They were everywhere, and not so coincidentally, they were one of the more popular fish of the period.

Advocates for the poor, pamphleteers, and local papers began calling for a cleanup as early as the 1820s, but to no avail. People in power would not do anything about it—not until it affected them personally. And it took the "Great Stink" of the summer of 1858 to make this happen.* During this hot and humid summer, merely waking from a night's sleep was enough to send

* For a more detailed description of this time and the man who changed it, read Stephen Halliday's excellent study, *The Great Stink of London: Sir Joseph Bazalgette and the Cleansing of the Victorian Metropolis.*

a person running to the nearest chamber pot to retch—thanks to the odors arising from the Thames and the city's streets. London's *City Press* declared, "Gentility of speech is at an end—it [the Thames] stinks; and whoso once inhales the stink can never forget it and can count himself lucky if he live to remember it." It was so malodorous that people thought they were dying from the vapors themselves. They fled the city in droves and the custodians of the Houses of Parliament—a new building along the Thames—took to dousing the venerable building's drapes with chloride of lime in an attempt to dampen the effects of the stinky air.

With this odor as the prevailing backdrop, Parliament was finally forced to begin debating the issue. Ironically, while so doing they were literally driven from their chambers by the stench. When they returned, with handkerchiefs pressed to their faces, they appointed Joseph Bazalgette, a noted civil engineer, to create an entirely new sanitation system, which eventually resulted in a series of drainage aqueducts that not only rivaled the Romans but actually surpassed them.

As previously alluded to, the existing sewers had merely been stopgap measures. Since they were originally conceived as drainage sewers of London, no thought had been given to their merit or responsibilities as disposers of waste. It was one of those situations where the people involved had simply tried to improve upon what was there, instead of rethinking and restructuring the entire mess. Bazalgette, the meticulous son of French immigrants, had submitted plans for a drainage system as early as 1853, after a cholera outbreak had killed more than ten thousand Londoners, but it took the Great Stink affecting the lawmakers themselves to coerce Parliament to break out the pocketbook.

Joseph Bazalgette not only devised and implemented an entirely new system that drained a city of two million people, but his plan also allowed for the city's inevitable growth. In fact, Bazalgette made the growth possible. He cleaned up the Thames, and thus much of London's water supply, by sending the sewage much farther downstream than ever before. He is credited with saving more lives than any other Victorian public official, mainly by realizing that the sewers, with outlets placed below high tide, did not allow the Thames to cleanse itself. At high tide, the drains could not discharge, keeping the stench wafting throughout the city, and at low tide, the sewage was dumped onto the shoreline, only to be swept slightly upstream with the returning tide and then back down to the city as the tide ebbed. The Thames was a splashtown of shit. (The same could be said for the Seine over in Paris, which underwent a similar experience during these decades. Unendurable stench, pollution, and cholera outbreaks also forced the French into action. The city of Paris hired an engineer in 1850 named Eugene Belgrand, who developed and built a system as unique as Bazalgette's. In my mind, the Paris sewer even has one up on London's because the sewer lines replicate the streets above. A wide street has a wide sewer beneath it and street signs can be found both above- and belowground.)

Work began on Bazalgette's system in 1859. It wasn't simply a matter of digging up streets and building modern aqueducts. Bazalgette and his workers were reinventing the wheel. He was a highly particular man who didn't leave anything to guesswork—even the bricks were subjected to constant quality checks and underlings were wary of his inspections. They created land where there had been none before, like the Victoria Embankment along the Thames where you can now hear summer concerts.

The project took decades to complete, but soon the stench was gone, cholera epidemics became a terror of the past, and the Thames, although not something you would want to swim in, was better, much better, than it had been. Bazalgette was the hero of his day.

(His sewers originally just moved the waste farther

downstream to two places: Beckton and Crossness. Fish certainly returned to London, but the fish, people, and communities downstream of London suffered greatly with this new concentration of untreated sewage. It wasn't until thirty years after work on the new sewers began that London started separating solids from liquids, dumping the solids into the ocean and the liquids, after being treated chemically, into the river. The first sludge barge in operation, the SS *Bazalgette,* went into service in the summer of 1887 and London continued dumping its sludge at sea until 1998, when England conformed to the European Union's ban on such practices. Meanwhile, Thames Water, the modern-day keepers of London's sewers and water supply, started selling electricity to England's national grid produced from burning methane at its Barking treatment plant. The remaining sludge is incinerated and the heat produced runs turbines which generate electricity that is also sold back to the grid. Today, with these more stringent sewage treatment rules in place, more than 115 species of fish are found in the Thames. Salmon and sea bass are caught right up to London Bridge and the Thames is considered one of the cleaner large metropolitan rivers in the world.)

Bazalgette's sewers are still in use today, and in May

2004 I decided to take a short stroll through his underground canals to see firsthand the wonders he had wrought. It would be a lark—a funny little tidbit to tell my future grandchildren: a tour of the London sewer. Each year in May, Thames Water, the private firm contracted by the London government, conducts tours of the sewer for two weeks, mostly taking down government dignitaries, employees, and the occasional oddball like me—and also my friend Rebecca Farley, who lives in England and has made it her part-time business to find plumbing-related items for me. As a university lecturer, she has plenty of time on her hands.

Arriving at the Abbey Mills pumping station—a facility built by Bazalgette to raise the sewage higher so that it could make it to the outfall station in East London—I was stunned by the detail lavished on this public works building. It's a gorgeous piece of Victorian architecture that rivals America's best town halls for sheer pomp and design. The Abbey Mills station was opened in 1868 and sits somewhat forlornly, like a forgotten child prodigy, in a frumpy area called West Ham in northeast London. Once the pumping station was online, the northern sewers were able to begin dumping in Barking—a few miles to the east. The southern system had already been in operation since 1865 and in 1875 the western system would be finished, making London's sewer and drainage one complete system for the first time. So it's not surprising that these buildings are refined structures—it's the magnitude of this refinement that gives pause.

The multistory brick building, nicknamed the Cathedral of Sewage, is topped with a gilded weathervane, the roof is covered with rich Welsh slate, and the gutters are fluted metal sculptures. The stone facade is carved with intricate floral details. Even the lead flashing has a delicate design. The door to the main pumping station stands twelve feet tall and is adorned with exquisite brass and copper hinges. Walk inside and it looks like a display case for designer ironworks.

When I asked one of my guides why it was so fancy, he replied, "They were ruling half the world. They were making a statement." Just like the Romans had been doing with their aqueducts and baths, saying, *See how much water we can control? See how much we can waste?* Except in the Britons' case, they weren't wasting the water so much as moving it out of sight and mind.

Back in Bazalgette's day, the pumping station employed 150 workers; today no one works there on a daily basis, except the guard—who, I noticed, drives an old Jag. Otherwise, the station runs itself, or rather, computers tell the pumps and gates what to do. (Actually, a new plant next to the "Cathedral" is where the pumping takes place today; the original plant is just a backup.) The pumping station sends nearly one million cubic meters of sewage a day to the Beckton treatment plant located in East London along the Thames, making the outflow from the plant the second biggest tributary of the Thames. So, maybe these buildings *should* be so impressive, considering the monumental amounts of waste they're in charge of.

Although the buildings were spectacular, I was most impressed with the wooden water pipe display inside the main building. My friend Rebecca, however, walked right by the two eight-foot elm pipes like they were uninteresting artwork, muttering, "That really ponked."

"Huh?"

"It ponked."

"What does that mean?"

"It's Australian. It means smelled. It really smelled like a sewer when we walked in."

"These are water pipes, Rebecca. Ancient relics. Wooden water pipes," I nearly yelled.

"Oh, that's a pipe?" she said, turning back. She touched the closest one. "You should make one of these, Hodding. No chance of getting poisoned doing this, huh?"

Ha, ha.

"How do you suppose they did it, bored them out?"

The display said they were bored out in the same manner that iron cannons were, which meant a lot if you knew how iron cannons were made. Later, I saw an illustration of a wooden pipe being bored with a tool that looked like an exceedingly long drill bit turned by a wooden windlass.

Oddly, communities around the world were still using wooden mains well into the twentieth century, even in the United States. The town of Conway, Arkansas, seat of Faulkner County, was a growing metropolis of around two thousand souls in the early 1900s, and as it lured in more citizens along with a couple of colleges, it decided it needed a public water supply. Wanting the best, it hired an out-of-state engineering firm based in St. Louis whose engineers went on a tour of the area, collected data, and chose a local waterway called Cadron Creek as the water supply. Then they suggested the town use wooden water pipes for their mains. The engineers claimed that wooden water pipes would make the water taste sweeter and would deliver the water in purer form—as compared to water delivered through pipes made of cast iron. (Of course, if they really wanted it sweet they could have used lead, but that would be a different story.) The engineers said the wooden pipes would have the same life expectancy as metal pipes, that acids and minerals in the water would not hurt the wood, and that wooden pipes would also be cheaper per foot, pointing out that wooden pipes with a ten-inch diameter would cost roughly sixty-five cents per foot, while cast-iron pipes would run more than twice that amount. Unlike old-style wooden pipes, these would be ironclad. A steel band would join them together and then the entire joint would be encased in asphalt. They were virtually leakproof.

Confronted with such facts and logic, the town fathers went with wood—cedar, to be more precise. In February 1911, a local contractor started laying down the pipe, and by the summer they'd already reached town. Unfortunately, many of the wooden pipes had already started to leak. Eventually, the town replaced all the cedar pipes with white pine and tried a new, supposedly leakproof method for binding the pipes together. The new pipes were expected to withstand pressures as great as 130 psi. However, one day in June 1914, a steel band popped off some pipes that crossed over the town's railway, nearly draining the reservoir dry, and turning Conway from wooden mains forever. A subsequent waterworks board meeting decided all the mains would be replaced with cast-iron pipes.

However, these two wooden pipes on display at Abbey Mills were magnificent examples of medieval engineering. Peering through the pipes, I saw there was a perfect seven-inch-diameter hole running their entire length. Where the two pipes met, the first was tapered so it could fit into the next. They leaked like sieves, but since elms and oaks covered England's hills, it apparently made a lot more sense than mining thousands of tons of lead. They used these wooden pipes as mains from the 1200s on into the 1800s in London and only used lead pipes as service lines from the mains. The wooden pipes always leaked at the connection, even when strengthened with iron bands, and so the system had no real pressure. Supposedly, it took hours to fill wooden barrels, even when they were placed at ground level. As a result, water was only supplied three days a week. The wooden water mains lasted anywhere from four to fourteen years, depending on the soil type where they were installed.

After our aboveground tour and a lunch of chili that tasted like Italian Bolognese sauce, we listened to a forty-five-minute lecture on the history of London's sewers. One thing repeated a number of times was the awesomeness of the brickwork—how it has held up for 150 years now and appears ready for another 150. And then, again, just before heading down, we were told the brickwork is the finest in the world—factoids mentioned, I guess, so that we would concentrate more on the bricks than on what was floating past.

Finally it was time to go below. They dressed the men in light blue throwaway jumpsuits and the women in yellow ones, and we all had weighted rubber hip waders to help us keep our footing in the slight current (it would normally be a swift current, but the flow had been diverted for the tour), white safety helmets with headlamps, rubber gloves, and emergency respirators. The last item was in case there was too heavy a buildup of gas in the sewer, which made me wonder how they used to work in the sewer back before electric lights. The answer was simple. They used candles and lanterns and sometimes they blew up—the lights and the people. Of course, it doesn't always take men with candles to make sewers blow up.

The Fleet River, which runs beneath Fleet Street, home of the British press, was once a clear running waterway, but like all other nineteenth century London streams, it was covered up and slowly became a sewer as the city grew. One day in 1846, the Fleet had had enough and simply exploded, destroying three houses as its fetid waters covered blocks of buildings and homes. Ever since, ventilation has been an important aspect of London's sewers.

Decked out in my safety gear, I felt pretty dorky and worried that this little sewer tour wasn't going to be adventurous enough. We weren't going on a quixotic exploration where we'd be hiking all day through the sewers and then finding some weir to sleep on at night—a weir constructed so that when stormwater rose too high it could spill over and fall directly into the Thames through various outflow channels.* We were just going on a fifteen-minute walk through a nearly pitch-black tunnel. Suddenly I realized I'd flown across the Atlantic for a stunt, and a fairly lame one at that. This was ludicrous. Or did I simply not want to wade through a stream of British poop?

It was too late to back out, however, so I gamely walked over to the manhole where they were lowering us down a ladder, with a winched cable attached to us to prevent our falling some twelve feet underground. How ridiculous, I thought, being belayed on a ladder. It was very dark down there and the water looked higher than I thought it was going to be.

"The water's pretty high," I told Rebecca. "It's probably going to knock you down."

"It's not water, Hodding," she said, smiling. Of our

*This overflow weir system is the downfall of the entire Bazalgette sewer system and what allows the Thames to be polluted today: the overflowing stormwater mixes with the sewage and they both engorge the river with enough bacteria to decimate fish populations. The meeting up or combining of storm drainage and sewage isn't just a London invention; it's the downfall of most big city systems all over the world. And it's why you never, ever want to go swimming the afternoon or even the day after a big rainstorm in any harbor, bay, or beach that is near a town or city. Never ever.

group of sixteen or so people, she was the only one smiling, except for one other guy who turned out to be a Ph.D. candidate writing a thesis on London's sewer system. All the other blokes worked for Thames Water and had to do the tour.

They hooked Rebecca up with her yellow suit puffed out over her waders, and then she carefully descended the ladder, not once looking back up to bid me farewell. A good thing, because then a droplet of fecal matter would have plopped into her eye and she'd have had ocular *E. coli* poisoning—a very hard thing to explain, I'd imagine.

"Remember, hold your breath from now on," I called to her, hoping she'd look up to laugh. Nothing. "Hey, Rebecca. Hold your breath. Rebecca." She never looked up but did punch me later when I got down there.

Finally it was my turn and I stepped onto the ladder, thinking how lazy it was and couldn't they at least let me rappel the twelve feet down. Then my foot slipped and I was falling into a bath of death—except the cable caught me and I regained my footing and shakily went lower and lower.

"Remember, don't look up!" one of the sewer workers, affectionately known as a "flusher" within the industry, called to me. The rungs were now dripping with sewer water—millions of pestilent microbes yearning to enter our bodies. They'd told us countless times not to look up and I'd just watched Rebecca carefully avoid doing so, no matter how many times I called to her, trying to trick her into it. Only an idiot would look up.

"What?" I said, looking up. A droplet burst apart on my right cheek, splashing within a millimeter of my eye. I was about to wipe when I realized how grimy my hands probably were and instead just looked down, hoping gravity would remedy the situation.

"Be careful," another flusher warned, detaching me from the cable and motioning me to walk forward a few feet to where the rest of the tour group was waiting. This one was our main guide and he looked a lot like the substitute ferry-boat captain that ran between Lincolnville and the island of Islesboro back in Maine. And, come to think of it, both men resembled some movie star, but who was it? Obviously, the odors of the sewer were getting to me. Did methane go for the brain first? "And remember, stay in the middle, because the floor is rounded and slippery."

Got it. Stay in the middle. Shuffle forward. Remember, 318 million bricks. Eighty-two miles of tunnels.

"These bricks have been here for a hundred and fifty years," the graying movie star was saying. "Quite a feat, to hold up this long. "

Scanning the bricked sides of the oval tunnel to show my appreciation for his appreciation, I saw something that looked like a high-water mark. It was at my eye level. I scanned it with my headlamp and then moved in close to touch it . . . Whoa! That was a close one. Both of my feet had slipped. I had stumbled and kicked up a splash in my efforts not to meet the queen's effluent head on.

"Steady, there," someone said all too pompously.

"Watch it," another squealed.

Okay, okay. The flusher was right. It was best to stay in the middle.

I was pretty cowed after that, content to walk directly behind Rebecca, the person in front of me. Since they had mostly cut off the flow to this canal, the water ran below my knees, but that meant that if you fell, your face would still end up underwater.

Suddenly a deafening roar drowned out all my little worries. My fellow spelewagers (a word I've just made up) were having a massive, collective freak-out. The place smelled like . . . sewage! All those quiet, stalwart souls were gasping and groaning in despair. The air was filled with "Ewws!," "Ahgghs!," screeches, and a fair bit of laughter. I asked Rebecca what she thought of the place so far, but I couldn't hear her answer over the din.

When things quieted down a bit, they lined us up single file, told us Beckton Works were four to five miles behind us and some other station was twelve miles ahead—I guess so that, in case of an emergency, we'd know to head back, not forward—and marched us forward.

As we walked along, a flusher said some more flattering things about the brickwork. It was hard to hear him over the splashing of my fellow spelewagers. The acoustics were overwhelming. The arched brick ceiling—arched just like a Roman aqueduct—sent his voice bouncing all over the place. So as he talked about the bricks, we trudged a few hundred feet. It wasn't as dark as you might think because every so often shafts of light descended through grated manholes. We eventually arrived at a juncture of sorts. Two tunnels that were mostly running parallel met up, and on the other side of two other weirs were two tunnels that were used for draining West London.

They took turns talking about the brickwork some more—Bazalgette would test the mortar himself—while they lowered a grated gate that is used to collect and extract large items flowing through the sewer—the usual stuff, like cats, dogs, carts, brooms, people, whatever. They'd also found more interesting things, like a live hand grenade, a bed, ducks, and a 150-ton

blob of cooking fat that took them eight weeks to remove. Movie-star flusher claimed that they never ran into any rats in the sewer, since rats don't live in water. Personally, I've seen nearly a hundred rats swimming over the span of my life so far, but I kept my mouth shut. If they wanted to believe that rats didn't live in the London sewers, then who was I to dispel such wishful thinking?

While we were standing there marveling at the gate, made of iron and still able to be lowered and raised after 150 years, I noticed a few floaters, well, float by.

"Oh, my," a lady in front of me said, and then started giggling.

"What?" her male companion asked. "You weren't expecting shit?" Suddenly ten beams of light were directed at the stream beneath us, catching dozens of floaters twirling past.

"Is that what I think it is?" another man asked.

Well, no, it's virtual shit e-mailed in so that you'll feel like you're really in a sewer, is what I wished I'd said at the time, but to be honest, I was just as surprised as they were. (I'm eating as I write this, so I know you can handle it. Just remember, it's a natural by-product of living and eating, and all animals, including us, do it.)

With the arrival of the poop, the tour didn't seem so stuntlike or, even worse, desultory. It felt refreshing.

Uplifting. I was even finding what the flusher was saying to be fascinating.

" . . . and the brickwork's mortar, as you can see, has really held up . . ."

Okay, that last bit may not be true, but now it was fun, at least. How often do you get to splash through the waste of millions of people? Hell, some of this shit might even be famous. That greenish brown one floating between Rebecca's legs might even be the queen's! The Abbey Mills line is part of the northern system, so it doesn't run beneath Buckingham Palace, but what the hell. She must use the bathroom—sorry, water closet—elsewhere sometimes. (Rebecca, who was born American, grew up in Australia, and now lives in Wales when not teaching in Portsmouth, says our fellow English speakers laugh at us for calling the toilet room a bathroom. How often do you find a bath in a restaurant's "bathroom"?) I suddenly realized I really might be cheek by jowl with royalty for the first time in my life. Not too many boys from Greenville, Mississippi, could say that.

A few minutes later, the gate came nearly all the way down, dripping with fecal matter, and then, after it rose back in place, they herded us to the ladder we'd originally descended. Walking back, I scraped my gloved hand along the side of the tunnel, like a kid will do along a fence with a stick, marveling at the number of people I was coming in contact with. This tunnel alone probably serviced some three million people a day.

Rebecca sent me an e-mail a few weeks after our adventure about the aftermath of recent rainstorms in London. Scientists and the press were blaming the antiquated sewer system for killing thousands of fish because the sewage and storm runoff blend together at high-water times. The excessive nutrients wasted the fish and so it seems that London, much like most big city water systems across the world, is in need of a reincarnated Bazalgette. He really did create a masterpiece for his time, but the sewers are ripe for something new.

7

It Makes Your Garden Grow

EVER NOTICED those egg-shaped battlements guarding the entrance to Boston Harbor? Those sparkling monoliths on Deer Island that can be spied from twenty miles out at sea? The first time I saw them I thought they were either navigational aides for sailors or some well-funded sculptor's ode to Mother Goose—an artistic declaration of Boston Harbor's rebirth, perhaps.

I was wrong on both counts. They're 140-foot-tall digesters—sewage treatment structures that kill harmful bacteria by feeding them to other bacteria—that are partially responsible for making Boston Harbor the lovely swimming hole that it is today.

Yes, in case you haven't heard, Boston Harbor is clean. Mike Dukakis can finally rest easy.

Boston has used Deer Island as a dumping ground for more than three hundred years. It began in 1675, when American colonists set up concentration camps for Native Americans during the King Philip War (King Philip was the English name for a native warrior called Metacom), where thousands of Indians died of starvation, exposure, and disease. Since then, it has also been home to a prison and a ward for mental patients. With this history, it should come as no surprise that Massachusetts is now using it to detain human waste. The good news, though, is that Deer Island is giving its current guests the royal treatment. The 390 million gallons of sewage that arrive at Deer Island on any given day—enough waste to fill up the city's eye-catching Prudential Building three times over—are treated to an experience rivaled by only a few other big city sewage plants in the world.

As many know, Boston Harbor used to be a scary place. With roughly two million people's untreated sewage dumped into it on a daily basis, this should have come as no surprise. It was a toxic bed of iniquity in which *E. coli* reproduced faster than bunny rabbits.

When George H. W. Bush campaigned against Michael Dukakis, he dubbed it the "Harbor of Shame." The water was so polluted that if you accidentally fell in, you were advised to rush to the hospital for a slew of shots and tests. Although the Clean Water Act, passed in 1972, forbade the polluting of local waters by municipal facilities, the Metropolitan District Commission—the Boston entity in charge of the area's two treatment plants—was granted waiver after waiver for more than a decade, forestalling what all hoped it would become: a cleaner, safer harbor. Instead, the harbor only became ever more disgusting.

Then one day in 1982, the inevitable happened: a bigwig got grossed out. Bill Golden, a Boston political power broker, stepped on human feces while jogging on the Wollaston Beach in Quincy.

Incensed, he sued the EPA, which in turn sued the Metropolitan District Commission. In 1985, the federal district court hearing the case ordered the state to clean up the harbor, giving specific instructions as to how it should be done and demanding that a new wastewater treatment facility be built that would adhere to the Clean Water Act. The state was given until 2000 to make the harbor both swimmable and fishable. The state then created something called the Massachusetts Water Resources Authority (MWRA), giving it authority to tax water users, and in 1989 the MWRA began constructing the $3.5 billion treatment facility at Deer Island.

The results have been astounding. First, and most obviously, you no longer die from disease when you fall into the harbor. Now it's one of the cleaner big city harbors in the U.S., and people come from all over to catch game fish like striped bass. All thanks to the digesters and other water-purifying equipment at Deer Island. At least this is what I'm told by the facility's public relations director. Personally, I can attest that I've gone swimming in Boston Harbor a number of times in the last few years and I feel fine.

Having experienced the results firsthand and wanting to see a big city treatment plant, I decided to take a look, heading there on the Friday afternoon of a July Fourth weekend—not the greatest timing considering Boston has some of the worst drivers in America, but somehow I made it unscathed.

From the outset, the place is not what I expected. Walking trails crisscross the island, interpretive signs dole out wildlife and historic trivia, and ospreys screech at one another just over your head. It looks like, and is, a park—part of the Boston Harbor Islands National Recreational Area that includes all of the local islands. It's a comfortable, inviting place. After a three-hour tour, the scariest thing I found were the "mangrates" located at the edge of the treatment facility.

"Those are meant to catch you if you fall in," Charlie Tyler, one of my guides and an engineer for the MWRA, explained. It was hard to hear him over the roar of falling water that was passing through the mangrate, a large metal screen that resembled a gate to a medieval castle more than anything else. Disturbingly, some of the grating was missing—eaten away by chemicals in the water—so there was at least a six-foot-wide gap. More than large enough for a person to slip through.

We were holding on to a railing, watching the cascading water. When I finally comprehended what he was saying, I stepped a few feet back. If I were to fall in, my next stop would be nine and a half miles out to sea, washed away in a turbulent stream of sewage—treated sewage, of course, but still sewage.

The treated water, plunging a hundred feet down before entering a twenty-four-foot-wide tunnel burrowed out of bedrock, sends up a sweet-smelling aroma. The day the secondary treatment facility went into operation, Judge Mazzone, the federal judge overseeing the harbor's cleanup, boasted that you could drink the stuff and held up a glass as if to do so. He didn't, and looking at the gray, murky water, I must say

I wouldn't, either—despite the process it must complete before being diffused into the ocean.

Speaking of which, this is what happens. The sewage comes to Deer Island from forty-three communities in the Boston area via four aqueducts, divided into a north system flow and a south system. Once at Deer Island, the flows are treated separately, because it was initially thought that the south-system poop would make better fertilizer since it comes from a less industrialized area, but this hasn't proven true. On an average day, 390 million gallons of sewage arrive at the island, but when there's heavy rain, this number can nearly triple since many Boston, Somerville, Cambridge, and Chelsea sewers also collect stormwater runoff—a no-no in modern city planning for obvious reasons. (Remember the fish kills in London.) The Deer Island facility can handle 1.2 billion gallons of sewage a day—a number that has only been reached once, during a really horrendous rainstorm, since the facility became operational. It was April Fool's Day, 2004. They couldn't treat everything coming through, so they "did some blending," which means they mixed completely treated wastewater with untreated wastewater and sent it out into the ocean, something that ideally should never happen.

Once at the island, the influent, as the waste is called, must get separated into two parts: liquids and solids.

That way the liquids can be purified and dumped out in the ocean and the solids can be made free of harmful bacteria. To begin this process, the influent is pumped 150 feet up, to the "head" of the facility, where it passes through grit chambers that remove heavy particles that can't be digested.

This machinery, loud and unwieldy, looked like it was straight out of a Dr. Seuss book. And it sounded like it, too.

Plink! Plink! Plunk! Plink! Plink! went the grit as it was spit out of dirty plastic tubes onto a conveyor belt that swept it out of sight and into a trailer that was emptied about once a week into a local landfill. The entire Deer Island facility is chock-full of air scrubbers and filters to keep the odor down but the degritting room stank as much as one might expect, since sewage-coated items were being tossed around in the open air. (The MWRA takes its stink seriously. Odiferous gasses are piped into a tower where sodium hypochlorite sprinkles down a silo, attacking the odor; from there the air is pumped through carbon filters, and only then is it released into the atmosphere. Deer Island even provides an "Odor Hotline" for residents of the neighboring town of Winthrop in case any objectionable odors do escape the facility.)

I moved in for a closer inspection of the grit and saw

a plethora of cigarette butts and pebbles, by far the most common items, followed by little sticks, nails, and some unidentifiable light brown button-shaped things.

"No one has been able to figure out what those are," Charlie said, when he saw me staring at them. "Some sort of leavings from industry, I guess." I was about to pick up one of the buttonlike objects when I remembered where it'd been.

After being degritted, the sewage flows into forty-eight clarifying chambers housed in what looks like a giant warehouse. To imagine how massive a space this is, realize that each chamber is the length of an Olympic-sized pool (though not as wide, at forty-one feet) and forty-feet deep. Here, the majority of the solids float to the bottom and are removed to the digesting area. There are layers and layers of these chambers, but since they're covered in concrete to cut down odor and are really just settling tanks, I discovered there wasn't much to see unless I was willing to sit in one of the tanks wearing a diving mask, something I wasn't too interested in doing.

From there, the liquid sewage flows to more clarifiers that remove more sludge, "floaters" (anything you can imagine—Styrofoam, condoms, straws, etc.), and other debris that didn't settle in the primary clarifying area. In these secondary clarifiers, pure oxygen (manufactured at Deer Island) is injected into the mix to activate microorganisms that consume organic matter so that the solids will continue to break down and settle.

The influent is almost ready to become outfluent. The liquid gets pumped into two five-hundred-foot-long tanks that can hold a combined four million gallons of fluid and is infused with sodium hypochlorite, the same stuff used by the air scrubbers. These tanks, covered only with metal grating, are located on the upper deck of the facility, just before the mangrate tunnel. The sewage, no longer stinky, is exposed to the open air for the first time since arriving at Deer Island. By this point, it smells like laundry detergent. Surprisingly—at least to me—the sewage still has solid objects in it. In something called a scum collection area I saw scum, naturally, but also tampon applicators, more straws, and coffee stirrers. How could these things get past the grit collection area and all the other screens? This is the last removal area, however. After these tanks, the water—it seems safe to call it that once again—spills over concrete walls and heads out into the ocean through the rotted mangrate.

Unsettlingly, given that it has left the treatment plant, the water is still not safe for the environment because of the large amount of chlorine used to disinfect the sewage. It would kill many marine creatures, creating a moonscape-like atmosphere on the ocean floor where the treated sewage is released. To keep this from occurring, the Water Authority must neutralize the chlorine in the water with sodium bisulfite—a sulfur also used in wine-making to prevent oxidation and preserve wine's flavor. But here's the catch: they've run out of space on Deer Island by the time the chlorine has done its job, so they inject the sodium bisulfite into the water as it travels through the tunnel. It's a major weak point in the system, seeing as it is out of sight, but by the time it reaches the fifty-one diffusers (water release areas) spread out along the last one and a half miles of the tunnel, the water is supposedly some of the safest-treated wastewater in the country. They do periodic checks to make sure the sodium bisulfite is working and there was a lot of sea life at the diffusers in recent photos I was shown.

But what's happening to all that sludge that's been removed in the primary and secondary treatment facilities? This, to me, was the most exciting part.

The sludge—which is a lot more liquid than you might imagine—is thickened up in centrifugal thickeners (tanks that spin around so that liquid gets "squeezed" out) and is then, still in a very liquidy state, pumped into the giant eggs. Finally, I was going to get to see the eggs.

"This is where we operators get excited, too," Charlie said. "All of a sudden, it becomes 'my sludge.'" Here, microorganisms found naturally in the sludge break down the remaining organic matter as the sludge is agitated and pumped around inside the eggs—very much like what happens in your stomach, except that the digesters are oxygen-free and are kept warm to speed up the anaerobic activity.

"If you have too many microorganisms and they run out of food—sludge—then you get cannibalism and you start over again. You've got to have exactly the right number of microorganisms," Charlie explained. "This perfect state is called a 'mixed liquor.' It's a British term; we get all our treatment terminology from them."

A little later, as we finally stood outside one of the eggs, he told me, "We're just trying to do what nature would do if it had enough space,"

Well, yes and no. Yes, nature would break down all this sludge if given enough room to do so, but no, the MWRA is actually taking it a step or two further. The microorganisms break the sludge down into sewer gas, carbon dioxide, water, and solid organic matter, and then separate them. Then the water gets sent for treatment, the carbon dioxide is released, the solids moved on to the next phase, and the methane-rich sewer gas is used to create power.

Yes, at Deer Island human waste is used as an energy

source. A large percentage of the sewer gas is captured and directed to the facility's power plant, where it is used to fire the boiler that produces hot water to heat the plant in colder months. But—and this is a big but in my mind—in the summer, when an excessive amount of gas is produced and hot water is not in high demand, most of the gas is flared off, burned in smokestacks before being released into the environment. This is a far better system than old treatment facilities that just released the gas, thus contributing to the deterioration of the ozone, but a far cry from perfection. They could actually be using all the methane all year long if they used the gas to power their generators—saving taxpayers money and helping the environment even more because they wouldn't be importing as much electricity made from burning fossil fuels.

I pointed this out.

"Yes, you're right," Charlie conceded. "It wouldn't be enough to make us completely self-sufficient, but it'd go a long way toward it."

"But we had to make an agreement with a local power company when we started operations here that we would buy the majority of our power from them," Jonathan Yeo added. (Jonathan, head of public affairs, and another man named Pat Costigan, who normally leads tours of Deer Island, had been with us the entire time but hadn't said much, since the operation of the treatment plant is Charlie's specialty.)

"Maybe you can do something different when that contract runs out," I suggested, and Jonathan nodded his head with a look that said, *Not likely.*

At this point, we were standing on a walkway that connected the tops of all one dozen of the eggs, 140 feet up in the air. It's the best view in Boston—in one direction you've got the outlying islands that look like scattered puzzle pieces slung across the ocean, and on the other you've got downtown Boston. It's a modern statement, this Fort Poop, overlooking, guarding, and creating a revitalized Boston Harbor.

A Plexiglas viewing window looked down into one egg and I was finally going to get to see where all the real action is—the intestines of the treatment plant.

I hurried over as Charlie pointed it out.

The egg was full of dark sludge and was not lit up by any lights. Dark sludge. Dark egg. Dirty, scratched Plexiglas. I couldn't see a thing.

Registering my disappointment, Jonathan said, "I don't know why they have these windows. They've never worked."

Just so I could see something, Charlie opened a hatch where the sludge enters the egg. It wasn't what I wanted—being predigested—but at least it was sewage. As one might expect, it looked pretty nasty and smelled like poop. I guess I was looking forward to seeing it bubbling and frothing. I'd imagined it as a cauldron of life—a witches' brew capable of creating great new worlds.

"Okay, so what happens with the remaining sludge—after it's been digested?" I asked, moving on, after the grate was replaced.

"Turned into fertilizer," Jonathan said. "It gets loaded onto a covered barge and then taken to a pelletizing facility on the mainland where it's dried up and made into a fertilizer." (Two fourteen-inch pipelines will soon be taking the sludge directly from Deer Island to the pelletizing plant.) "It's marketed as Bay State Fertilizer and used all over—from Boston Common to citrus farms in Florida. . . . It's pretty good stuff. You can get it at local nurseries."

"Do *you* use it?" I couldn't help asking.

"Yeah. It's great," he answered. Charlie and Pat kept mum. "Although it does have a faint odor after the first rain."

8

Blame It on the Christians

A "FAINT ODOR"? He meant the compost smelled like feces, right?

Why can't we say that?

Eliminating our bodies' waste is a natural, daily function. Why do we have to beat around the bush with euphemisms and vague references to what we're talking about?

From now on, look your host directly in the eye and say, "Where is your toilet? I have to urinate (or defecate)." Please don't ask for the bathroom. What, are you going to take a bath or shower in the middle of a dinner party? I don't think so. You're going to urinate. Be honest.

The euphemisms of this century for the room containing the toilet are forever increasing, and not only are they incomprehensible to speakers of other languages, they don't even make sense to people in different English-speaking lands. And when they do make sense, they're downright silly. Here's a handful from this century alone: bog (is it that smelly?), can (is it big enough?), cloakroom (aw, no, not on the coats), dump tank (a new defense department boondoggle), dunny (dunno how that one started, mate, but watch out for that dingo!), the facilities (what are you? a visiting field inspector?), the gentlemen's room (and if you're not one?), the john (are you a prostitute?), the ladies' room (you're a woman, not a lady, remember?), lavatory (no way—not in my sink), little girls' room (pervert!), the necessary (who says so?), the place of easement (I actually like this one), the powder room (okay, but let's hop into bed right afterward), restroom (like I said, I've got a bed), the shit-house (that's not a euphemism; it's just plain gross), the shitter (not bad, but still, it'll stink), the smallest room (not anymore), the throne (okay, but only if you bow down), the porcelain god (ditto), the washroom (why do you need to wash? come on, tell me), the head (why do sailors have to come up with new names for things that already have names?), the water chamber (that's why we have glasses and a tap in the kitchen), and the potty (okie-dokie, baby-boo-boo).

And then there are the ones for the actual action of defecating or urinating: take a leak, talk to the president, attend to pressing matters, take care of business, make a short (or long) call, crap, pee, poop, piddle, peewee, dookie, wee-wee, tee-tee, doo-doo, tinkle, whiz, relieve myself, see a man about a horse . . .

You might say we make up all these words and phrases because we're embarrassed about the odor. But do feces and urine really stink or have we merely convinced ourselves (or been convinced) they do? How did we get so squeamish about odor and about feces, urine, and the process of eliminating them from our body?

The privatizing of urinating and defecating may not be the root of all fears and anxiety, although I'd argue the case, but it has certainly played a starring role.

"You'd be surprised how often this comes up. . . . Well, *you* wouldn't, maybe, but normal people would," a psychologist friend of mine told me when I confronted her with the subject outside her urine and feces room. "It's discussed nearly every day in my practice and it's such a devastating concern. People don't want others to know they've gone to the bathroom. They worry about people hearing them when they're in there. They're even embarrassed that it smells. And they're often very afraid to talk about it."

Sound familiar? Ever run the water?

Is this any way to live? Who would you rather be—the Roman, farting, pooping, and peeing just inches from her best buddy without a twinge of guilt or embarrassment, or one of us, wetting our pants with fear that someone might think our shit stinks? It's ludicrous. Yet it's so pervasive you probably can't even see my point. We think it completely natural that one would want to hide the fact that he has defecated. Our culture, and most others these days, demands it of us.

I can personally testify as to how similar feelings were reinforced for me. The big sport among the boys, besides freeze tag and S.P.U.D., in my second-grade class at Carrie Stern Elementary School in Greenville, Mississippi, was to torment whoever was using the toilet. During the "bathroom" break, after using the urinals, we'd kick open the swinging door of each toilet cubicle, taunt the innocent defecator with "Poopy head, poopy head," and other such witty jibes, and then, if the victim remained seated, toss soggies at him. Using the toilet was already a shameful, private matter, as we all knew, but this put it into the realm of That-Which-Must-Not-Be-Done. Things got so bad that I quit using the toilet at school and tried my best to hold it in while walking the mile home. To this day, as a result, I try to be as quiet and innocuous as possible when sitting on a public toilet. (These days, as a blossoming bathing, urinating, and defecating—BUD—zealot, I don't exhibit such reserve when using the toilets at home. I make a point of leaving the door open when I'm on the toilet, inviting my kids in to talk about their day. As a result, we gather around whenever anybody is using the toilet, continue conversations that started in the kitchen, and generally hold forth as if we're yakking over a bowl of popcorn. This will either warp them for life—their mother's belief—or liberate them, unleashing a brave new generation of BUD activists on the world.)

Not convinced it's such a big deal?

One of the functions of a humorist or comedian is to lampoon or pop the bubble of cultural mores or taboos. And one of the satirist's favorite subjects, from Rabelais all the way down to Larry the Cable Guy, has always been bathroom humor. It's the biggest target out there and they've been blasting away at it for five hundred years or more, simply to bring some psychic relief—letting us out of the closet at least for that cathartic moment when they imitate someone farting or by recounting a silly scatological experience. Here's Gargantua telling his father Grangousier the best way to wipe after defecating in Rabelais's *Gargantua and Pantagruel*:

> "I have," answered Gargantua, "by a long and curious experience, found out a means to wipe my bum, the most lordly, the most excellent, and the most convenient that ever was seen."
>
> "What is that?" said Grangousier, "how is it?"
>
> "I will tell you by-and-by," said Gargantua. "Once I did wipe me with a gentlewoman's velvet mask, and found it to be good; for the softness of the silk was very voluptuous and pleasant to my fundament. Another time with one of their hoods, and in like manner that was comfortable. At another time with a lady's neckerchief, and after that I wiped me with some earpieces of hers made of crimson satin, but there was such a number of golden spangles in them (turdy round things, a pox take them) that they fetched away all the skin of my tail with a vengeance. Now I wish St. Antony's fire burn the bum-gut of the goldsmith that made them, and of her that wore them! This hurt I cured by wiping myself with a page's cap, garnished with a feather after the Switzers' fashion.
>
> "Afterwards, in dunging behind a bush, I found a March-cat, and with it I wiped my breech, but her claws were so sharp that they scratched and exulcerated all my

perinee. Of this I recovered the next morning thereafter, by wiping myself with my mother's gloves, of a most excellent perfume and scent of the Arabian Benin. After that I wiped me with sage, with fennel, with anet, with marjoram, with roses, with gourd-leaves, with beets, with colewort, with leaves of the vine-tree, with mallows, wool-blade, which is a tail-scarlet, with lettuce, and with spinach leaves. All this did very great good to my leg. Then with mercury, with parsley, with nettles, with comfrey, but that gave me the bloody flux of Lombardy, which I healed by wiping me with my baguette. Then I wiped my tail in the sheets, in the coverlet, in the curtains, with a cushion, with arras hangings, with a green carpet, with a table-cloth, with a napkin, with a handkerchief, with a combing-cloth; in all which I found more pleasure than do the mangy dogs when you rub them."

"Yea, but, which torchecul did you find to be the best?" said Grangousier.

Eventually, Gargantuan comes to the point:

"Afterwards I wiped my tail with a hen, with a cock, with a pullet, with a calf's skin, with a hare, with a pigeon, with a cormorant, with an attorney's bag, with a montero, with a coif, with a falconer's lure. But, to con-

> clude, I say and maintain, that of all torcheculs, arsewisps, bumfodders, tail-napkins, bunghole cleansers, and wipe-breeches, there is none in the world comparable to the neck of a goose that is well downed, if you hold her head betwixt your legs. And believe me therein upon mine honor, for you will thereby feel in your nockhole a most wonderful pleasure, both in regard of the softness of the said down and of the temperate heat of the goose, which is easily communicated to the bum-gut and the rest the inwards, in so far as to come even to the regions of the heart and brains."

Makes you happy that somebody invented toilet paper, originally called boudoir paper, in the early nineteenth century.

What brought on this type of humor? What made us so conflicted and uptight about such a basic, universal experience?

I blame it on the Christians. In the Middle Ages, the Church implemented changes in plumbing and enforced new thinking about BUD that have altered our collective Western psyches to this very day.

In this era, BUD—practices that had been public activities, at least among one's own sex, in the past—became totally private matters in much of Western Europe. Suddenly these things were only allowed in darkened alleys, behind closed doors or wooden partitions, in a pot, in an enclosed cubicle, or quietly behind a bush. (A bit later, during the Renaissance, some royals boldly relieved themselves while in company on ornately decorated apparatuses called close stools, but these were relatively short-lived fads.)

As the Christians took control of Rome and then filled the power vacuum throughout Europe, they inherited some incredible water systems. In many instances they maintained them when it came to piping in water to their monasteries and abbeys, and in some cases they even improved upon them. But they also made a few changes as to *how* they were used that created all this mess I'm talking about.

While the commonfolk were usually mucking about behind bushes and toting their water from God knows where, the church filled its own cisterns—cylindrical structures made of brick or stone and often lined with water-impervious material such as cement—via sometimes miles-long lead and wooden pipes. From the cisterns, the lead pipes, fashioned in the same manner as the Roman variety, ran underground to different parts of the living quarters, called dorters or dormitories. Many of these monasteries had sanitary wings called rere-dorters, used for defecating and urinating, that were separated from the main dormitory by a bridge or elevated walkway. These latrines were often located on an upper floor, perhaps to make it easier for removing the contents via an underlying stream.

It is within the structure and layout of these latrines that our current BUD practices began to take shape. The monastic privies were very similar to the Roman latrines in that they were all lined up together and were often made of stone, on which the users presumably sat. Sometimes the seats were placed back to back, even. The crucial difference between these facilities and the Roman ones were that these Christian latrines were private. While placed extremely close together, the monastic latrines always had some sort of partition or wall between them. In "A Description or Brief Declaration of all the Ancient Monuments, Rites, and Customes Belongine or Being within the Monastical Church of Durham before the Suppression," written in 1593 about life in the medieval monasteries, we're told there was a large house close to the water called "the Privies" and that "every seate and particion was of wainscot, close of either side, verie decent, so that one of them could not see one another, when they were in that place."

The message was clear: if you're going to poop, it's going to be in private.

Unsurprisingly, this era gave rise to a new word,

privy (derived from the Latin *privates,* meaning set apart), to mean a latrine. John Barbour, an archdeacon of St. Machar's Church in Scotland, was the first to use it, in his epic poem, *The Actes and Life of the Most Victorious Conqueror, Robert Bruce King of Scotland,* written in the mid-1300s: "The King had in custom ay, / For to riss airly eurilk day, / And pas weill fer fra his menze, / Quhen he vald pas to the preve." (The Bruce is King Robert the Bruce, Scotland's first true national hero—as opposed to that William Wallace fellow that Mel Gibson made us all fall in love with—who defeated the British at the Battle of Bannockburn in 1314, winning Scotland's independence.)

But why make it private? The simple explanation is that Christianity had to dominate and perhaps wipe out the practices, mores, and customs of the religions it was replacing, and as it spread, it was replacing all sorts of different religions: Roman multitheism, animism, perhaps druidism, and Norse gods, to name a few. Historically, when one religion takes over another's congregation, they make taboo that which the other worshipped, adored, or even simply permitted. Most of those previously mentioned religions apparently did not forbid public defecating and urinating and some clearly encouraged it, as in the Roman piss pots placed on street corners or outside fullers' shops and in its many communal latrines where people gathered to gossip and conduct business while relieving themselves. Many of these religions also considered human excrement as symbolic of the beginnings of life. For example, the Romans revered Sterculius, god of odor, because he was the first to lay dung upon the earth.

To combat the influences of Sterculius and the like, and to set itself apart, Christianity declared natural bodily functions abhorrent and chose to demonize those moments when bodies shed their waste. Furthermore, this secrecy might also have been a holdover from a Judaic edict spelled out in Deuteronomy 23: 12–14. Moses told the Israelites that when they went off to war they must "set up a place outside the camp to be used as a toilet area. And make sure that you have a small shovel in your equipment. When you go out to the toilet area, use the shovel to dig a hole. Then, after you relieve yourself, bury the waste in the hole. You must keep your camp clean of filthy and disgusting things. The Lord is always present in your camp, ready to rescue you and give you victory over your enemies. But if he sees something disgusting in your camp, he may turn around and leave." From this, the Christians learned that human waste is disgusting, would lead to the Lord's disapproval, and perhaps that it should be done in private, since Moses directed them to defecate outside the camp—although this was clearly a matter of hygiene rather than privacy.

Christian saints actively encouraged people not to bathe. Saint Benedict commanded that healthy people never wash and one of precocious Saint Agnes's virtues apparently was that she had never washed. Saint Francis of Assisi, best known these days for the prayer "Lord make me an instrument of thy peace," believed grubbier people were more holy.

This abhorrence of bathing was also a reaction to the practices and beliefs of the supplanted cultures and religions. The Romans, as we have seen, glorified in public bathing—originally only in same-sex situations but later, as its power waned, in mixed-sex ones as well.

Saint Boniface, a British-born believer and highly accomplished destroyer of idols and pagan temples, forbade such bathing in 745 and the baths themselves were labeled *seminaria veneata,* hotbeds of vice, by the Christians of his day. Saint Boniface, by the way, also serves as a useful example of how many of the edicts or actions of the church were merely done to replace the old religions. While converting souls in Saxony, he happened upon a group of people worshipping a Norse god who at this particular site was represented by a beautiful six-foot-wide oak tree. Boniface must have been a fairly burly fellow because at this point he walked up to the tree with the crowd gathered around him, took off his tunic, picked up his ax, and started chopping the tree down. He didn't stop until the tree had crashed to the ground, whereupon he stood on the remaining trunk and boasted, "How stands your mighty god now? My god is stronger than he." Evidently, some of the crowd weren't too pleased with Boniface's actions and contemplated chopping him down as well, but others immediately converted to Christianity.

Not all Christians felt as Boniface did, though. Pope Adrian I, 772–795, not only recommended weekly bathing but oversaw improvements to Rome's water system. And Saint Francis of Assisi referred to aqua as "our Sister Water, very serviceable and humble and precious and clean."

Also, as we've seen, monasteries were equipped with running water for both bathing and privies. The message then was hypocritical. The religious leaders could bathe and enjoy the benefits of good hygiene, but not the common people. A typical example of this hypocrisy in action was to be found at Christchurch Monastery in Canterbury in the mid-twelfth century. Water from a local source filled a large tower and then ran downhill via an underground lead pipe through five settling tanks and then back to the monastery. It was collected in another tank, called a conduit, which was raised on stilts aboveground to give the water enough "head" to branch off to various parts of the monastery. "Head" is pressure created by raising water to a certain height so that it may reach its terminus. The different supply lines ran to lavatories in the scullery, kitchen, bakehouse, brewhouse, guesthouse, infirmary, and the bathhouse and remained constantly running, just like in the Roman days. Another line also filled a conduit for the townspeople to gather water from. Drains were built that ran from the bathhouse, the prior's rooms, the roofs, and then under the reredorter, taking the wastewater out of the monastery's grounds. Unsurprisingly, the Canterbury monastery escaped the devastating Black Plague in the fourteenth century and thus it becomes much clearer why Chaucer's Wife of Bath and her fellow pilgrims were heading for Canterbury. It was probably the cleanest place in England.

Despite these exceptions and perhaps because of the success of such establishments as Canterbury's Christchurch, bathing, urinating, and defecating grew ever more private. And in many ways, this privacy—the practicing of it in close quarters—is what brought about the development of the toilet as we know it today.

9

The Birth of the Toilet

As we have seen, most medieval Europeans did not have water systems akin to those used by the monasteries. They had to urinate and defecate in chamber pots, garderobes, and close stools, and as a result they were left with a great stench in their homes. An entirely new type of privy was needed.

Sir John Harington, godson of Queen Elizabeth, developed something he called an ajax—a play on words (*jakes* was slang for a privy at the time)—with several friends while gathered together in 1594 at Wardour Castle. An accomplished writer and translator, he was by no means an inventor or plumber but had a firm grasp on common sense—judging by his ajax and his writing. In 1596, he published *A New Discourse on a Stale Subject Called the Metamorphosis of Ajax,* complete with drawings of his device and specific instructions on how to build one.

Clearly, Harington believed his invention worthwhile and important for others to duplicate, offering to help anyone who wished to build it: "I will come home to his house to him. . . . I will instruct his workeman, I will give him plots and models, and do him all the service I can: for that is a man of my own humor." Yet, he wrote under the pseudonym Miscamos, not only because the subject was so base, but also because the pamphlet was a dancing, jabbing, weaving attack on his courtly contemporaries—a bit of satire meant to provoke and entertain as much as educate. While most of his puns and ripostes are lost on modern readers (I would have been completely lost if not for Elizabeth Story Donno's annotated version of 1962), Harington's satire was a big hit in Elizabethan London. There were many printings and even pirated copies circulating among the well read, even though he did upset his benefactor, the queen. While he was off on a 1598 campaign against Ireland, a cousin wrote to him: "Your book is almost forgiven and forgotten; but not for its lacke of wit or satyr . . . tho' her Highnesse signified displeasure in outwarde sorte, yet did she like the marrowe of your booke." The queen, in fact, liked the "marrowe" of his work so greatly that she had Harington build her one of his ajaxes at her palace in Richmond.

In the beginning of his pamphlet, Harington contends that the slang *jakes* refers to the Greek Captain Ajax, who upon his suicide morphed into grass. What happened, according to Harington, was that one day a young gentleman told his servant to mow some hyacinth so he could use it in the privy. The owners of the grass told the gentleman that the grass was of the ancient House of Ajax and should not be used. The man used it and was struck with Saint Anthony's fire in the posterior. He then went on pilgrimage around the world for a cure and after finding it made a vow to honor Ajax—that he would much rather use a piece of paper from Holinshed's *Chronicles of England, Scotlande, and Irelande* than Ajax's grass. (Raphael Holinshed's works were very popular during this era and were the source Shakespeare plumbed to write *Macbeth* and *King Lear,* among others. Harington evidently found Holinshed intolerable.) When he returned home, he built a privy and hung a statue of Ajax over the door "with so grim a countenance, that the aspect of it being full of terrour, was halfe as good as a suppository," and called the place Ajax. Harington then explains that it was eventually garbled into "a jakes."

Perhaps to entertain people into taking his invention seriously, he fills most of the pages with witty barbs against other scholars and many of his contemporaries in Elizabeth's court. Or perhaps the entire invention is a joke and is simply a device for writing a funny book on such a private matter—as it was a very private matter by this point in history.

Here, you decide.

"He that makes his belly his God, I wold have him make a Jakes his chappell," Harington wrote. "But he that would indeed call to mind, how Arius, that notable and famous, or rather infamous hereticke, came to his miserable end upon a Jakes; might take just occasion even at that homely businesse, to have godly thoughts; rather than as some have wonton, or most have, idle."

Either way, he does eventually describe how to make the ajax, providing plans, materials list, and illustrations. Here are the written instructions:

> In the Privie that annoyes you, first cause a Cesterne containing a barrel or upward, to be placed either behind the seat, or in any place, either in the room, or above it, from whence the water may by a small pype of leade of an inch be convayed under the seate in the hinder part therof (but quite out of sight) to which pype you must have a Cocke or a washer to yeeld water

> with some pretie strength, when you would let it in.
>
> Next make a vessel of an oval formne, as broad at the bottome as at the top, ii foote deep, one foote broad, xvi inches long, place this verie close to your seat, like the pot of a close stool, let the oval incline [decline, in modern usage] to the right hand.

Enough. I read the entire directions a dozen times and still didn't understand them. He also provides a part list with prices, followed by a drawing for the workman to see the parts, which is in turn followed by a drawing that shows the device put together, with each part marked and annotated. The drawings make the whole thing understandable. I also like the explanation of the seat in the parts list the most, which details a "peke devaunt" for elbow room, which turns out to be a "short beard trimmed to a pretty polywigge sparrows tayle peake." In other words, the seat should have a little outcropping so there's room for your penis (if you're a man) to hang down.

The vault into which the waste falls is also shown, and he reminds the reader "always remember that at noone and at night, emptie it and leave it halfe a foote deepe in fayre water. And this being well done, and orderly kept, your worst privie may be as sweet as your best chamber."

A few important details were left out—like what should the stopple be made of—but the system did work. Since the next breakthrough in privy technology did not occur for a few centuries, many toilet historians have made the mistake of claiming that Harington's ajax never caught on and that no one had a water-run privy for two hundred years, but that wasn't true, as there are references to ajax-like privies in both England and Europe after the publication of Harington's pamphlet.

The part that didn't catch on was the name *ajax.* Evidently, it was replaced with the words *water closet.* A Sir William Hamilton presented a lecture on archaeological digs in Pompeii to the British Antiquarian Society in the late 1700s that confirms this: "Closest to the Temple of Isis is a theatre, no more of which has been cleared than the scene and corridor which leads to the seats. In the corridor was a retiring place for necessary occasions, where the pipe to convey the water, and the bason [basin] like that of our water closets still remain, the wood of the seat only having mouldered away by time."

Other references to water closets also make it clear that Harington's ajax, under the name of water closet, did indeed catch on. In 1770, the city government of Bath, England, warned a Mr. Melmoth that they would cut off his "pipe unless he discontinued the supply to his water closet."

These water closets were nowhere near universal and most people continued to empty their pots directly into the nearest street or cesspool, but Harington's ajax did help ensure that bathing, urinating, and defecating remained private.

Close stools, essentially chamber pots masquerading as seats and other items, remained the rage. Practically anything you could think of to disguise the stools' true purpose was used—from a box that looked like a stack of books to a trunk to an ornately carved seat. The volume-of-books model was used in France and the fake books were given titles like *Voyage au Pays Bas* or *Mystères de Paris*.

Slowly, the close stool was augmented with the pan closet, which consisted of a hinged pan being placed under the bowl or seat you sat upon. When you were done, you pulled a handle that tipped the pan down, the contents dropping into some container or, on rare occasions, perhaps piped into a nearby cesspit. This was still a very messy and odorous affair. The pan's seal was not very good and it had to be cleaned constantly.

In 1775 a watchmaker named Alexander Cummings came to the rescue. The same profession that solved the longitude problem also got the stink out of our houses or allowed us to come in from the cold—depending whether you used a close stool or an outhouse. Cummings applied for the first patent in England (or anywhere) for a valve closet, being a water closet with a valve, or some sort of flaplike device, located at the bottom of a metal bowl. A handle was pulled or moved, which allowed water to flood in as the valve slid out of place at the bottom of the bowl. Once all the water flushed through the bottom of the bowl, the valve slid back in place. To make sure the stink stayed put, he also connected an S-shaped pipe to the bottom of the bowl that worked as a siphon. This latter part of his invention—the S-trap—however, was evidently not new. Although there is no written record of S-traps being in existence before this time, Cummings wrote in his patent request that such a trap was "too well known to require a description here." However, with the combination of the water-sealed valve and the S-trap, humans were finally on their way to getting the stench out of the closet and thus the house—as long as you were using a cesspit with its own vent.

Cummings's revolutionary device was quickly improved upon by cabinetmaker Joseph Bramah, who applied for a valve closet patent of his own just three years later. Bramah had had the luck of installing a number of Cummings's designs and came up with a more trustworthy version of the valve. As both installer and inventor, he quickly outpaced Cummings and within eleven years had installed six thousand of his own valve closets. The Bramah and a number of ripoffs would become the main design in valved water closets for the next sixty years.

The French installed valve closets, either direct ripoffs of these men's designs or actual Bramahs themselves, during this period. Contemporary letters and artifacts make it clear that the French acknowledged that the British were the ones who came up with the working water closet, as the new-style privies were called "English conveniences." Parisians who wanted to set apart rental apartments were sure to mention the modern pad in 1790 as having "*un grand cabinet de toilette à l'anglaise.*" This is also where the word *toilet,* meaning a water closet, sprang into existence. Originally, a toilette was a dressing table and all its accoutrements, including a *toile* covering that hung to the floor. It was originally used as a euphemism for the terms *water closet* or *lavatory,* taking hold in England and the U.S. in the twentieth century, although it was considered a somewhat impolite word in the U.S. until very recently.

Although Bramah's numbers were staggering and steadily grew, these early water closets did not catch on wholesale; most people stuck with their close stools and pan closets. The valve closets were too expensive, as each was individually manufactured, and so were status symbols—sort of like having a Jasmin washlet today. They also required an abundance of water as well as somewhere for this increased waste to drain. As we have seen, houses and buildings weren't connected (or weren't supposed to be) to the sewer system, which was still just for storm drainage. So, if you were going to have a Bramah, you had to have enough land to have an even bigger cesspit than ever before. Across Europe, the valve closet was a popular luxury in the early 1800s—not a necessary (that would soon become a euphemism for a water closet as well) fixture.

Meanwhile, in the U.S., it was neither a necessary nor a luxury. It simply didn't exist. America, still trying to shrug off England, lagged far behind in what would become known as the sanitary arts. In fact, America has remained in this secondary position to this day. Any major innovation in American plumbing—whether it be tools or techniques—has come directly from Europe. It started with the water closet and continues up to this past year with the introduction of PermaLynx, a new system for connecting copper pipes without soldering. The same technology, hailed as the biggest breakthrough in American plumbing since the plunger, has been available in Europe since the 1980s.

Back to the late 1700s and early 1800s. If Americans had anything at all, they had outhouses—only a very few innovators had actual privies. Thomas Jefferson, for example, probably had three of the earliest indoor privies in the United States installed at Monticello, but thanks to thoughtless renovations performed on the house in the 1950s that used the privy shafts for installing air-conditioning and heating ducts, there is no remaining evidence of how these privies worked. What is known is that the privies had a seat of some sort and that a shaft took the waste down a few feet below the basement. Jefferson referred to them as "air closets," which might imply that they were not water-driven privies. The three shafts emptied into a masonry-lined conduit (also called a "sink" by Jefferson) that was 2 ½ feet by 3 ¾ feet running at a 3-inch decline every 10 feet. The conduit then ran for about 125 feet, terminating east of the house.

According to the folks currently maintaining Monticello, Jefferson had used water closets while he lived in Paris and thus might have wanted a similar setup at home. It's hard to know, since Jefferson only made a few oblique references to the closets. He wrote some instructions to a joiner working on them: "I would much rather have the 2nd and 3rd air-closets finished before anything else, because it will be very disagreeable working in them after even one of them begins to be in use." Given this reference to what one might suppose is the stench of sewage and the labeling of the basement conduit as a sink, then it does stand to reason that these were privies and that they might have been operated by water.

Jefferson's indoor privies were more the elite oddity than the norm, according to contemporaneous accounts. Visitors to the fledgling United States often complained of the lack of English conveniences and the awfulness of having to go outdoors to relieve themselves. A Frenchman named Moreau de St. Méry can be counted in that number. A leader of the French Revolution, Moreau barely escaped the guillotine by fleeing on a ship bound for America when his political opponents took power in 1793. Within a year he set up shop in Philadelphia and later wrote an account of that experience that belies our distorted image of early life in America. In *American Journey,* he complains, "Each house has for a toilet a small room set apart from the house, but it is far away. One often gets wet going to it." Moreau actually has very little good to say about anything or anybody in America. American men, he says are tall, thin, listless, and have no strength. American women fare even worse in his estimation: "After eighteen years old they lose their charms, and fade. Their breasts, never large, already have vanished." Worse, though, is that American women felt conflicted about their bodies and according to Moreau they thought Frenchwomen were "unclean" because they dressed in ways that showed their chemises. "I am ashamed to say that it is exactly because American women are so sensitive about these garments," Moreau relates, "and because they have so few of them and change them so seldom, that they are guilty of not keeping them clean, and of dirtying them with marks of that need to which Nature has subjected every animal." Evidently, American women were not using corncobs or goose necks but their undergarments, something that Moreau frowned upon. But it does bring up an interesting question. Why do we wear undergarments? These days, when we wash our clothes after each wearing, the undergarments clearly serve no purpose—except for brassieres, of course. At some point, however, especially in light of Moreau's accusations regarding American women's undergarments, weren't they worn for the express purpose that Moreau frowns upon? People weren't really using goose necks despite Gargantua's claims, and clearly most people did not have a handy barrel of corn, especially in England and Europe, where corn didn't exist until after the colonization of the Americas. They used their undergarments. Undergarments could be washed out at the end of each day, and they couldn't be seen. What else could they have been for?

Be that as it may, water closets did not catch on in the U.S. for some time. A perusal of catalogs, broadsheets, and newspaper ads from this period gives no evidence of the sale of water closets in America until the 1820s, and no one applied for an American water closet patent until 1835, and that was for a portable water closet, essentially a close stool. Portable WCs were popular because they did not require plumbing, much like today's Porta Potties. All you needed was a cesspit or a local sewer or an old outhouse. Over in England, a company called Wiss had a number of different styles, which were throwbacks to the old close stool days. Some looked like hard-back chairs, others like a set of bed steps, and another like a side table. They were all quite tricked out with a four-gallon "cistern," pump, "air trap," and flushing basin.

But as indoor plumbing advanced, water closet usage increased exponentially over the next half century, with Americans originally simply importing British designs. While most American houses in 1850 had only a hand pump and a sink either in or adjacent to the kitchen, more and more were installing lead pipes throughout the house to service sinks, baths, and water closets in various parts of the house. The key to this development was a pressing desire to reform the American character, according to Maureen Ogle, author of *All the Modern Conveniences*. Towns and cities were rife with disease, crime, and "social evils." Civic leaders fixed on water-works as the key to salvation—increasing the number of communities with water works from only 50 in 1840 to 240 by 1870. Meanwhile, the average townfolk turned to fixing themselves as well, trying everything from religion to vegetarianism to hydropathy—the last of which required cold water immersions, foot baths, douches, and the like. Architects got in on the movement and based all their designs on the magic word of the day: convenience. And what was more convenient than having indoor plumbing and any number of water closets? Architects and how-to-improve-your-life guide writers soon started labeling the water closets themselves as conveniences.

"At mid-century Americans began to install the water closet, tub, and pump, as well as the dumbwaiter, speaking tube, and furnace, as part of a contemporary effort to reform and improve American domestic life," Ogle writes, "and it is within the dual contexts of reform and convenience that the introduction of plumbing is best understood."

(There is something a little disturbing in the use of

the word *convenience* as a euphemism for the water closet. In the eighteenth century, *convenience* was slang for a mistress, after a while giving way to meaning a wife. How'd it become a euphemism or slang for a water closet? What was happening in those early water closets?)

At first, Americans could only import British water closets, specifically Bramah's or similar knockoffs. Because only a few Americans homeowners wanted or could afford indoor plumbing, this was not considered a hardship, but as America began to prosper and New England grew more industrialized, plumbing and toilets became increasingly popular. From the 1850s on, American plumbers and inventors applied for an ever-increasing number of patents relating to toilets and plumbing, closely following but never preceding British innovations.

Meanwhile, the British, ever improving their invention, finally came up with the revolutionary design: George Jennings patented a valveless water closet in 1852, forever getting rid of the annoying valve—what had euphemistically become known as a "dirt trap." Jennings was a crusader of sorts, vociferously arguing for the construction of what he called "Halting Stations" all over England wherever people assembled. He offered to supply and fix the water closet halting stations for free and man the booths with attendants if allowed to charge a nominal fee. He wrote, "I know the subject is a peculiar one, and very difficult to handle, but no false delicacy ought to prevent immediate attention being given to matters affecting the health and comfort of the thousands who daily throng the thoroughfares of your City." However, they initially refused him, not giving the go-ahead for public water closets until the 1870s.

While Jennings's toilets worked, they didn't always work perfectly, often failing to flush away all the matter and frequently overflowing. In fact, one never really knew for certain which way the waste might go: down the drain or onto your new pair of rubber-heeled shoes (the new rage). Also, there was the inescapable problem of odor. These water closets often had vent pipes, which took the horrible fumes out of the house, but the entire systems weren't properly ventilated like they are today. The deadly odors from the water closet pipes were leaving the water closet rooms, but they weren't necessarily leaving the building. The gases simply recirculated into the home through the waste pipes of the sinks, baths, and showers. Eventually, some genius—and it's not known who it was and whether he was British or American or perhaps even French—realized that venting the entire plumbing system would not only rid the home of noxious odors but also create enough air pressure in the system to allow the siphon of the water closets to work properly. And that's how you got that black pipe that sticks up out of your roof today.

Once this problem of ventilation was solved, indoor plumbing, specifically water closets, was the next big thing. Sanitation engineers grew insistent that water closets not only be used and that sewer systems take the waste as far from cities as possible, but that plumbing be seen as the savior of civilization. S. S. Hellyer, a water closet inventor and owner of a plumbing manufacturing firm in England, perhaps best summed up the fervor of the moment in a speech to London's plumbers: "Lying there in those strong arms of yours, slumbering in the hardened muscles, resting in the well-trained fingers and educated hands, lies the health of this leviathan city. The plumber's part in making a house, a town, a city healthy is enormous. Let the plumbing be done on the principles laid down in these lectures and this huge city, teeming though it be with human beings, will become the healthiest, as it is now the greatest, city in the world."

Perhaps as a result of such exaltations, the second half of the nineteenth century was a confusing time of toilet proliferation, with inventors and manufactures each declaring their product as the final breakthrough.

Sometimes it was the valve closet, sometimes the improved hopper, other times the valveless closet. But in 1870, a T. W. Twyford, of Hanley, England, brought out an all-earthenware closet called the washout, in which the trap was built into the entire water closet. The pan and valve closets had always kept the bowl separate from the working parts beneath, creating great stench and leakage. This all-in-one design was the birth of the modern toilet. As far as the basic concept goes, the washout is what the British and Europeans still use today. The waste in the bowl is pushed out by the flooding of the bowl. Around the same time, a similar water closet was introduced that operated on siphonic action. In this model, the waste is sucked out of the bowl. This is the style of toilet used in North America, Canada, and Asia.

Since then, American and European inventors merely improved upon the same design. The biggest change occurred in the early twentieth century when the water tank was moved from high above on the wall to resting on the back of the bowl. Everything else was mere tinkering.

The real work was being done by the men (and a few women) who kept the toilets functioning: the plumbers.

10

"What Was in the Toilet?"

When I ask a modern plumber what's the most important thing a person needs to know about plumbing or how to be a good plumber, he'll answer, "Shit runs downhill and payday is on Friday." He says it quickly, followed by a derisive chuckle, invariably grimacing at the tiredness of this professional homily. Of course, it's true, but there's also so much more—from the ins and outs of civil and mechanical engineering to electricity, pipefitting, and all aspects of water supply.

Plumbers and certain civil engineers, unlike the rest of us, know how our entire water system works. After Armageddon, a nuclear holocaust, or whatever disaster might befall us, it's the plumbers who'll be able to provide their fellow survivors with the one thing they will need to survive beyond a few days: water. The plumbers among us will be our knights in droopy jeans.

I also like to ask plumbers why they became plumbers. Now, I'm sure he's out there, but I still haven't come across the man or woman who answers that a plumber is what he or she has always wanted to be. As much as I want to hear it, no one has told me he had sat in seventh grade social studies daydreaming about the day he'd get to do the rough-in (the laying of pipes and drainage while a house is under construction) all on his own. Many plumbers fall into the profession by chance. Someone in town needs an extra hand, a teenager gets hired, and soon he's on his way to being a bona fide apprentice. The rest are simply born with a silver wrench in their hands: their dads were plumbers and so, by God, are they. America's most recognizable plumber, Richard Trethewey of the television show *This Old House,* is a fourth-generation plumber. His great-grandfather Harry Trethewey came over to Nova Scotia from Cornwall to work in the copper and tin mines but soon decided he needed a fresh start in the New World. At the turn of the twentieth century, plumbing was an exploding business in America, much like Web search engines are today, and Harry and his brother got in on the action early, setting up shop as the Trethewey Brothers in 1902 in Boston. It's been the family business ever since. "I like to joke that I used to get wrenches for Christmas, but my parents always said I had a choice," Richard told me. "As the oldest son of a fourth generation business, however, there was never much doubt what I'd be doing." At age twelve he started working on holidays and summer vacations and hasn't stopped since. "I started at the very bottom—sweeping floors," he recalled.

Like many trades, plumbing has a strict hierarchy—levels that can only be attained after a certain number of years and/or hours on the job. There is no nation-wide regulation of this (or of anything having to do with plumbing, actually) and so different states require different things. In general, though, state codes are similar and this is how a career in plumbing goes from the bottom up:

You get hired as a trainee by a local plumber. Not having any pertinent skills, you're paid anywhere from minimum wage to $12 an hour. You have to do a lot of the errand work—running to the store for three more brass couplings—and backbreaking labor, like the digging I did in my basement. You're the lucky guy who gets to fix the broken sewage line or unplug the stopped-up toilet or snake out the drainage. While you might be making better money than your buddy at McDonald's, you get no benefits.

If you are diligent and show promise, after about a year you become an apprentice. You may not get much of a pay raise, but in many companies you'll get put on the full-time payroll, which means benefits: vacation and sick time, a 401(k), medical insurance, and maybe even school tuition. George Haselton, who runs a plumbing company in my hometown of Rockport, Maine, even gives his employees incentives to exercise. George is an avid surfer and exerciser. He figures he's at his best when he's in shape and that an in-shape workforce is a happier and healthier workforce, so he pays for health club memberships and gives paid time off for exercise.

Besides a dental plan, you also get a good deal of autonomy as an apprentice, once you've shown a certain amount of competence. You might plumb an entire house under construction, gluing about a mile of PVC pipe in the process, or you might remodel an older home's waste system, but you'll always have someone above you, checking out your work. There are state codes for about every single joint a plumber might make and it's the apprentice's job to know and follow all of them, or at least the ones that the local code enforcement officer cares about.

After you've been apprenticing for about a year and a half, you can become a journeyman plumber if you pass the state-certified exam. (Since the early 1900s, wannabe plumbers have had the chance to get their first years out of the way by going to plumbing school, earning state-certified degrees at community colleges and trade schools. Taking this route, you usually finish school in a year and come out a journeyman.)

At this point, as the name journeyman implies, you get some confidence-boosting independence and a significant pay raise. As a certified journeyman, you can be hired anywhere in the U.S. or Canada. You can ride the rails with tools in hand: have wrench, will travel. A friend of mine worked for a bunch of plumbers from Mississippi who crisscrossed the U.S. plying their trade—and did it by only plumbing Holiday Inns.

While your boss bills you out at $80 an hour, you might net only a little more than half that in wages, depending on your company's overhead and benefit package. As far as responsibilities go, it's equivalent to being on the midlevel management side of things in the white-collar world and financially you're doing a lot better than most of the T-shirted world of freelance writers, photographers, and artists. A quick search on the Internet shows job opportunities for journeymen plumbers from California to Canada's Baffin Island, way up in the Arctic Circle. Salaries range from $25,000 a year to $55,000, with housing included in more than a few cases. The Baffin Island job, by the way, doesn't *require* fluency in Inuktitut, the Inuit language, but does say that it would be an asset.

As a journeyman, your pay is good and job opportunities are everywhere. However, you still can't go out on your own; state agencies would shut you down faster than dog hair clogs up a bathtub drain. It takes five years of work as a journeyman to move up to the top of the field: the master plumber. By then, if you're still working for someone else's business, you might be making $40 an hour. If you want, you could go out on your own and bill clients well over a hundred dollars an hour in big cities. Of course, then you'd have to get your own insurance, hire someone to take care of your books, rent an office and storage space, buy a van, and eventually hire an apprentice so you can take care of all the paperwork.

Since plumbing emerged as a profession in the U.S., around the mid-1800s, that's how it has worked. You put in your hours and years and in time you get to call the shots and get paid good money doing it. Which brings up a bit of controversy. Why such shock and outrage at how much a plumber gets paid? Is this because we don't think their work is very difficult or takes much training? Perhaps, but I think the issue is a little more complicated. It could be that some people don't like paying a plumber very much because, putting it simply, they handle our shit. Or maybe they think people doing dirty or disgusting jobs don't deserve to be well compensated. But that's not always been the case—gongfermors, those medieval England cesspool emptiers, made two or three times the average worker's wages.

I think there are two main reasons people resent plumbers' bills. The first is that plumbers come into our homes fairly frequently. Carpenters and electricians and maybe even masons also enter our homes, but it's usually a one-shot deal. They build or install a certain item and are not seen again for years. Not so the plumber. He's always fixing some mess we've created. People fear and, therefore, mock their plumbers for literally knowing what we're made of. Plumbers are like Sidney Poitier in *Guess Who's Coming to Dinner*. They might be charming, gorgeous, erudite, but we're not going to like them—at least we think we're not going to. Mainly, it's because they're there and they're seeing it all. Whether we want them to or not and whether they want to or not, they see the dust in the corners, the dirty underwear under the bed, and the letter from the bank saying we're overdrawn again.

They see us without our game face and without makeup.

Second, not only do plumbers come into our homes, but they enter our most sacred and intimate places. The place many of us don't even want other people to know we use—as if. When I'm using a public restroom, say in an airport, I'm as quiet as shy kindergartner on the first day of school, hoping that no matter what transpires, the person sitting on the toilet next to me has no idea I'm there. Some people I know don't even let their friends use their master bathrooms. But the plumber goes in there often—sometimes way too often.

And most damning of all, because of the nature of their work, plumbers see and smell our waste. Thus, they know the one thing we're all trying to hide from the rest of the world with our perfumes, perfectly coiffed hair, pressed jeans, and country club memberships. They know we're only human.

Even if you're a billionaire corporate master-of-the-universe type, you still have to take a shit and your plumber knows it. You might drive around in a flashy red German convertible, laughing at the peasants walking down the street, but your plumber knows the truth about you. He knows that in the end, you're just one of us.

My friend George Haselton, the surfing plumber, knows this kind of truth about approximately a third of the people in my town, since he owns one of the three large local plumbing companies. With his swimmer's build and college education, George confounds the stereotype of the plumber. This goes for how he got into the trade in the first place back in 1975. "It was either that or go to jail," he says. "I was hopping freight trains in Canada having the time of my life when I got busted. They said I could either go to jail or join the military. Soon afterwards, I was in Air Force boot camp in Texas."

That's the first version he gave me, which, while partially true, left out some of the better details. It wasn't until a year later when he was adjusting sinks and showers for a summer resident on a local island that I got the whole story. George was taking the flow restrictors out of the sinks and showers—"No one is gonna tell these guys how much water they can use," he intoned mock-seriously—when he let me in on a little more of the story. After graduating from high school in Montclair, New Jersey, in the spring of 1975, he decided to hitchhike across country with his girlfriend. Things weren't exactly right between them and George was convinced the girlfriend wanted to be with George's best friend. When they reached California, he found out he was wrong. She wanted to be with another girl. Somehow, George ended up living in a "hippie commune thing" for the summer.

When it was time to go back home, he attempted to hitchhike, was robbed, and then decided to hobo it by riding the rails up in Canada. Things went wrong there as well. He sort of got involved in a cigarette heist—he was the unwitting lookout—with the police on his tail. A few days later, while the train he'd jumped was riding out through the prairies—"It was one of the most beautiful things I'd ever seen"—the Mounties caught up to him. Back in court in Kamloops, B.C., the judge realized George's grandfather was a well-known New York lawyer and let George off with a warning and enough money to get back home. Also part of the deal was that George would enter the military, something he'd been planning to do since graduation, when he got back home.

George signed up for the Air Force and by August he was at Lakeland Air Force Base in San Antonio, Texas.

And that's almost how George became a plumber. By this point in his story, George had the shower head back on after removing the flow restrictor. Most states now require the use of flow restrictors in any new installations but given this is an island and the water is coming from the homeowner's own well, it doesn't really matter that George is taking it off. "I can see why this guy wasn't liking this thing, being a Republican. It's clearly a Liberal-Democrat shower,"

George explained. "It's got protection and restrictions everywhere."

George took a military aptitude test to see where he should be placed and scored in the ninety-ninth percentile in engineering. They wanted to make him a nuclear weapons mechanic. "I was a hippie. I told them I couldn't do it. That I'd be a security risk—not believing in the use of nuclear weapons. So they said I could be honorably discharged. I said I still wanted to serve and they said I could learn a trade—a lowly thing in their minds. They gave me ten choices of what else I could be. My top choice was hydraulics. The rest were a bunch of building trades. My last choice was plumbing. That's what I got."

After training, George did two years of nighttime house calls at Vandenberg Air Force Base in California. The work wasn't very exciting, but it was better than fighting, and it was there he acquired one of his best plumbing stories. One evening, he was called into the home of a full colonel. The colonel's wife told him the toilet was stopped-up. Back then, he found it easiest to unclog a really stopped-up toilet by taking the bowl (the base that you sit on) out in the yard because the toilet had to be turned on its side so he could get into the concealed S-trap. So, as usual, he carried the entire thing onto the lawn and, reaching up through the base of the trap, he immediately found the culprit: a giant carrot carved in the shape of a penis. He covered it up and put the toilet back together in the bathroom. George was about to leave without saying anything when the colonel asked George what had been the problem.

"Son, son, wha'd you get?" he asked. "What was in the toilet?"

"Sir, have you been away a lot?" George asked, and then handed over the homemade dildo. The colonel looked at it, shook his head, and marveled, "That's huge!" Then he offered George a drink.

At the end of his military stint, George went to the University of California at Santa Barbara and then led the enviable life of an itinerant plumber with his wife, Eliza. Surfed the California coast. Fished salmon in Alaska. Delivered sailboats to Africa and Europe. Cruised the Caribbean. Whenever money ran low or funds were needed for the next adventure, he'd do some plumbing. In 1990, he and Eliza decided to settle down in our town. George got his Maine license and opened up Haselton Plumbing and Heating. Initially the plumbing business existed only to pay the bills; in the summer, he captained boats for summer vacationers, saving the long winters for plumbing and installing heating systems.

While he still takes time to sail, surf, and do just about anything else that involves playing in the outdoors, he's turned his sideline profession into a highly successful business, renaming it Rockport Mechanical. He plumbs for the likes of John Travolta at his summer home on a nearby island in Penobscot Bay and countless year-round dwellers. One day he might be giving a talk on geothermal heating, and the next he's off to Cape Verde for a week of surfing. In 1999, he and his family spent a year in Kauai so he could ride the big stuff.

George let me tag along with him and his crew this past year. At the time, he had twelve full-time workers—a number that often seems to fluctuate. George spent most of his time checking on their work, doing estimates for future work, fixing leaky pipes at established customers' homes, or just offering free plumbing advice to friends seeing that he's a walking-talking-plumbing reference tool for his friends and acquaintances. While we were driving to see a friend who needed help installing her $6,000 French oven, a surfing buddy called from California and George suddenly sounded like one of the Click and Clack guys on the National Public Radio show *Car Talk*.

"What's the matter, your head still hurting from that last wave?" George asked when his friend said things weren't going too well.

His friend's well pump was malfunctioning. The motor worked, but nothing was coming up.

"Oh, that's easy. It's just a clog. You've got to ream out all the orifices," he said. "Get a coat hanger out of the closet. . . . Yeah, one of the wire ones. Stretch it out and poke it in all the holes. Flush them out. That ought to do it. Of course, since I'm not there, I don't know, but give it a try."

The friend thanked George.

"And if that doesn't fix it, work harder in the store, then hire a plumber," George joked just before hanging up.

He knows plumbing like the rest of us know our navels. At one point, we stopped in a favorite eating place in Camden, an old drugstore called Boynton McKay that serves wraps and burritos. After we ordered, he dragged me into the bathroom with him.

"We've gotta inspect it. Just can't help it," he said by way of explanation. "I do it wherever I go. Second nature."

It's a tiny bathroom, barely big enough for my daughter Helen, let alone George, who, I've neglected to mention, is six-foot-three and all arms, but I went in anyway, eager to learn.

"Look at that," he said, not indicating anything as far as I could tell. When he realized I didn't know what the hell he was talking about, he touched the metal radiator that sat adjacent to and below the toilet bowl. Much of the yellowing paint on top had chipped away and the radiator was starting to rust. The cause? Urine splashing off the rim of the toilet bowl.

"I always look for urine splash when I go into a public toilet. Old habit. . . ."

"Is urine splash a really bad thing?"

"Naw, just corrosive."

"This bathroom's got pretty typical problems," he went on. He pointed to the sink. "That S-trap under there? It's against code. Illegal. Can't put them in anymore. It siphons the waste back up. Could make it flow right out of the sewage line and fill the sink. Awful." Remember? That's why we all have P-traps now.

"That faucet? It's a cheap Taiwanese thing. It'll be leaking like a sieve in a couple of months." He spun around, accidentally pushing me off balance. Someone was knocking on the door.

"The toilet's fine, though—low gallons per flush. Overall, this isn't so bad. I give it a passing mark."

My chicken fajita wrap was good, too, but then it always is.

Another day, we drove out to inspect the motor on a faulty well pump at a home fifteen minutes outside of town. It was a cold autumn morning—the fallen leaves were iced with frozen dew—and the homeowners were worried not only about their pump, but also about their filtration system, which appeared to be clogging up; the heater, which wasn't always kicking on; and a drainage problem in the basement. After a few minutes' inspection, he explained to the homeowner what was wrong (or not) with each item and set about fixing the things he could do right away.

The most illuminating problem was with the pump. Something was wrong with the wiring. He could fix it back at the shop in just a few hours—charging the client a little more than $100 for his time—or buy a new part and replace it for less. "The problem with plumbing in America is that the industry is dumbed down," George told me. "What I mean is, labor is so expensive that it's become cheaper and easier to replace an item than fix it. Got a broken hot water heater? Throw it away. Busted faucet? Buy a new one. The most obvious downside to this is fuller landfills. That sucks, but it's just the way it is. It's different over in Europe. They still take time to fix things. Make them last longer."

I told George how most people think plumbers are just ripping them off. He nodded—heard it all, I guess. I then told him my theory of why plumbers don't get much respect—because we're embarrassed about how they know the details of our private lives.

"Yeah, we know details even we don't want to know. But, man, do I know what you're talking about," George added. "I've always thought that they got it wrong with Jesus. They made him a carpenter to show how humble he was, right? Well, they should have made him a plumber."

George went on to tell me about the Buddha phallus that he once fished out of a toilet, the male mayor in Utah wearing a sheer teddy whom he encountered during a late-night emergency house call, and the two women who didn't stop having sex when he walked in on them in broad daylight at a local bed-and-breakfast.

When George tired of me hanging around and asking him questions, he introduced me to some of his younger plumbers, Dewey and Mike. They trudged up the short flight of stairs to the white-walled office set over George's garage, joked about something from the day before, and did just what George hoped they wouldn't: hang out. Assignments were posted on a dry-erase whiteboard, but Dewey and Mike's names didn't have anything next to them.

"You're gonna have some fun today. I got a call yesterday evening. These people need their poop cleaned out and you and Dewey are the lucky contestants," George told Mike. "I checked it out last night.

Shouldn't be too difficult, just a backed-up clean-out, I think. The basement is flooded a bit, though."

Dewey, a skinny twenty-three-year-old transplant from Lafayette, Louisiana, smiled. This was Dewey's second week. He'd married a local girl and moved up here the previous year. Mike, twenty-eight, a similarly skinny twenty-eight-year-old from Camden, appeared to be grimacing. Mike had been at Rockport Mechanical for a couple of years.

"There are no turds or anything like that floating around," George continued, "but it is a sewer line. The original clean-out wasn't done quite right, so working the snake is going to be a bit tricky." George explained how to do it and Mike listened intently, asking questions, while Dewey and I waited in the background.

On the way out the door, as if to relieve some tension, Dewey joked, "It's gonna be a shitty situation." I rode in Dewey's pickup. Since Mike had been with the company longer, he had a Rockport Mechanical van: white, no logo.

The lack of logo or any advertisement for Rockport Mechanical is intentional. They're not even listed in the white pages. "I feel like I'd rather be able to select our customers and it's fun to be a stealth company," George explained, sneaking a smile, when I asked him about it later. "I guess it's a kind of reverse snob thing. If people really want us, they'll find us. Or maybe I'm just cheap—don't want to pay for the listing.

"Seriously, in a small town, if you're not getting work by word of mouth, then you're already making a mistake," he continued. "We're a part of the community; we live here. If I make a mistake, people know where to find me."

So Mike drove the unmarked van while Dewey and I took his truck. "We don't do these kinds of things very often," Dewey said on the way over, referring to the basement problem. However, each day that I went out with someone from Rockport Mechanical, somebody was dealing with a plugged-up toilet or stopped-up septic system or a similar situation. Yet, each time, I was told they don't get jobs like that very often. As Dewey got a call from his wife reminding him he forgot to leave his paycheck at home, I wondered why there's this level of denial even in the profession itself. Yes, it's probably true that most of their time is not spent on jobs like this. The majority of their work is plumbing new supply and waste lines and installing heating systems for new houses. That's where they spend most of their hours. Also, these "shitty situations," as Dewey calls them, are relatively easy. They usually take less than an hour. But I think there's also a desire, even by plumbers, to distance themselves from the fouler side of their job.

The flooded basement was at a modest summerhouse on Camden's Chestnut Street, a tree-lined avenue running the length of the high land overlooking the south side of the postcard harbor, although that fall nobody was taking pictures. The boats had been cleared out early so years of sediment and debris could be dredged out of the inner harbor. The muck being loaded onto the dredge barge was releasing a colossal stench, easily overpowering anything the basement might be dishing out. Years of sea-life decay and the illegal emptying of boats' holding tanks had built up quite a large amount of trapped hydrogen sulfide and methane. Picture-perfect Camden smelled like a stinky mudflat, which, if you've never experienced such, smells like a sewer. (In summer 2004, the town had to put up signs telling people not to swim in the harbor because the *E. coli* count was too high. Since then I've heard stories from local carpenters who have seen broken cast-iron sewer lines emptying directly into the harbor. So maybe we weren't only smelling the results of natural decay and a few boats' wastewater.)

The Chestnut Street homeowner kindly asked us if we wanted some coffee as she showed us to the basement, which had an aboveground entrance in the back since the house was built on a hillside. The real plumbers declined the offer, so I did, too.

As George promised, stinky sewage greeted us at the entrance, but it was very low and confined to a relatively small area. A tall man with a southern accent, a man we'll call Steve, walked in just as we were setting up the snake through the clean-out pipe.

"If you see some orange pee, it's my wife's," he told us without more than a how-do-you-do. "The medicine she's taking is making it that way. Just wanted you to know in case you thought it was weird or something." He was a good bit older than us—mid-sixties—and got friendly blank stares in return.

"Any you boys married?" Dewey and I nodded. Mike had a long-term girlfriend and future marriage plans. "Then you've dealt with this, right?"

Not yet. By this time, Mike had started up the motor for the snake and it was crawling through the pipe when suddenly there was a slight sucking noise over the din of the machine. The sewage was ebbing.

"Glad I went the right way," Mike said. It'd been hard to tell which way to feed the snake, not knowing exactly where the clog was.

"Yeah, me, too," Steve said, imagining the snake winding its way up to the toilet. "That would have been a real shocker to my wife." As we cleaned up, more of the household popped in to say hi, all dressed in their bathrobes. Plumbers see more people in their nightclothes in a week than the rest of us will see in a lifetime, but these people definitely didn't care.

Throughout it all, Steve stuck around, and finally it became clear why. He wanted to share his favorite plumbing story. One thing I learned while making the rounds with George's plumbers: people like to wax scatological when given the chance. If you don't believe me, try bringing up the subject during your next dinner party. I guarantee the poop stories will flow, even in the most prudish of crowds. (Okay, maybe not at dinner. Try it during cocktails.)

Steve told us that he and his wife lived in South America for a while and the particular city they lived in had a problem with snakes in the sewers. "You'd be sitting on the toilet and all of a sudden this large snake would come up and poke you. Talk about startling. Rats, too. Made you not want to use the bathroom. . . . Hope you don't mind my hanging out watching and talking so much, but I have to tell you I'm a frustrated plumber. I was a chemical engineer, but I just love this stuff."

The next day, I tagged along with Mike to a home under construction where he'd been working. All the

PEX supply lines had been stapled in place and the waste pipes were glued together and stood gleaming-white between the studs like stalwart white knights. It looked like he was about done until we climbed a ladder into the attic. There, a labyrinth of plastic pipes crisscrossed the space. Most of them were just lying on the floor, near where they were going to be glued into position.

The most important thing when working with PVC is to make sure of your measurements. Check and recheck them and don't glue anything together until you've already cut all your pieces and put them together without glue. "My best guy, Brice, does it slowly and methodically," George told me one day. "When I hired him, his old boss said he was a good worker but really slow. Really slow. This was back in '91 when we were first starting up. I took him on a job with me. We were working on a line from opposite ends. I was working my butt off fast and hard. I keep looking over at Brice and I don't see anything. I'm thinking, man, this definitely isn't going to work. By lunch, I've got a fair bit of PVC laid—probably halfway there. Brice doesn't have a single joint glued together. We go back and a couple hours later he's done and I'm still nowhere near finishing."

Mike was clearly of the Brice school of plumbing:

measure, cut, lay out, and then glue. Surely, though, all this pipe couldn't just be for venting?

"Anything that has a drain needs a vent. Maybe a condensate drain from an air conditioner doesn't, but otherwise, all those drains need a vertical pipe taking the smell away," Mike said when I questioned him.

We went to work immediately and after a while I was feeling a little bit happier than usual. The label said not to use the glue in an enclosed space, but of course that's where plumbers always are. The attic had already been insulated and there was just a small hole in the roof where Mike was going to place the flange for the vent stack. (A flange is just a plastic or metal plate that has a hole in the middle for the pipe to go through; a stack is the line of a drainage, waste, or vent system that all the other pipes fit into.) At the time, I was really, really liking those words: *flange, stack, flange stack, flange-stack.*

I asked Mike if he was feeling the same. "Yeah," he said, barely nodding. It was not a bad feeling and I could see why so many homeless kids resort to sniffing glue. I was feeling pretty good, really. Suddenly I didn't mind that I didn't stop wetting my bed until I was twelve or that I was not going to be a movie star. And the pipes were looking beautiful. So white. So perfect. But it turned out there was a real reason for that.

"We like to hide the lettering on the pipes. It makes for a cleaner display," Mike said, as he rubbed some pipe cleaner on the outside of the PVC to erase all the red and black markings that the manufacturer had inked all over the pipe. The rest of the pipes had the parts with writing on them turned to the wall.

Modern plumbing is a lily-white world.

I remarked that he seemed to be taking a lot of precautions for something as seemingly insignificant as venting. It's not like it's high-pressure water pipes, I said. But, of course, it is just as important. As already explained, sewer gases can kill you. Proper venting is what moved plumbing from a pipe dream into reality starting in the late 1800s. Without it, a lot of us would either be dead or blown up. Obliterated by our own waste. I suddenly found myself worrying about a vent pipe that I'd installed in my own remodeled bathroom. I hadn't been able to get the glue on for some lame reason in a half-covered wall, so I had just pushed the pipe in place, thinking the tight fit of the PVC connection would be enough. Mike told me venting has to be able to hold 5 psi for fifteen minutes to pass code inspection and added, "You can test with water and they do in some places, but you have to fill the whole thing, and if there's a leak, you're screwed." I made a mental note to go back and tear out the wall that took me two days to build and glue the vent stack together.

Before becoming a plumber, Mike worked as a carpenter for a local contractor, so I asked him what he liked more: pounding a hammer or gluing pipes. "I like plumbing much more. Most of the time with carpentry, you're stuck outside. We get more indoor jobs. Call me a wimp, I guess, but I like this more."

I pushed for a more competitive answer. "Which one takes more skill?"

"Plumbing," he said without missing a glue joint. "In carpentry, if something is slightly out of whack, you just take a sledgehammer to it to make it plumb. Not so with plumbing. You'd crack it apart. You have to get it exactly right at the outset or redo the whole thing. Plus, there's a lot more codes for plumbing. You have to be a master plumber to legally plumb a whole house. Not for carpentry—anybody can build somebody else a house."

My time spent with George and his crew did not make me a much better plumber, I'm afraid, but I did walk away with a number of useful facts—glue starts to harden at ten seconds, don't put the radiator too close to the toilet, never flush a dildo, improperly set ballcocks can make a toilet tank continuously fill and thus lead to condensation on the outside of the tank that can leak down to the floor and make the johnny bolts rust apart, leading to a rocking toilet bowl and big Uncle Bert lying on Italian-tiled floor awash in all that stuff we don't like to think about.

And that's that.

Except for this last word from Mike, after I asked him the hardest thing about his job: "It's kind of confusing when the homeowners talk to you about how to fix the problem, or even what the problem is, because they don't really know the system—how it got here, where's it's all going, and how it all fits together."

But, of course, now you do.

11

Touching the Untouchables

WHAT DO YOU do when you have no plumber, or even indoor plumbing, to turn to? If you're on a camping trip, you buy a copy of *How to Shit in the Woods* and study the suggested techniques. Then, while on said camping trip, you grab a trowel and toilet paper and sneak off deep into the forest. You find a hidden spot fifty feet from any water source, scan every direction for the fourth time to make sure nobody can see, and then, when satisfied with your privacy, you dig a few inches beneath the surface and take care of your business, perhaps frequently swatting at mosquitoes taking advantage of your disadvantage. (I've actually almost fallen into my hole while doing such a thing, so beware). You bury what you've left behind and march back to camp, proud of your handiwork and perhaps trailing a short length of toilet paper that got caught somewhere in your haste.

If, however, this is an everyday situation for you because you live without an adequate water supply or wastewater removal, you adopt a more casual approach and simply go in a pot, behind a bush or car, or wherever you can. That's what an estimated two billion people still do today, roughly seven hundred million of them in India alone.

In 2004, only 232 Indian towns out of 5,003 had sewer systems. The remaining 4,771 towns' citizens mostly use dry latrines, or, more often, nothing at all. A dry latrine is any situation where humans defecate into a bucket, basket, or any other receptacle that is not connected to a sewer system. According to statistics provided by Sulabh International, an India-based nonprofit organization created to uplift the untouchables and build more sanitary latrines throughout the country, India still has ten million dry latrines and employs some six hundred thousand untouchables to empty them, despite a 1993 law explicitly banning the use of such latrines as well as the employment of untouchables to manually carry human waste. The untouchables sit at the bottom of the Hindu caste system and are the only people deemed low enough to handle human waste.

Why are they "untouchable"? Because they carry and handle everyone else's feces by either emptying chamber pots and buckets or shoveling out those dry latrines. Often they have to scrape the waste up off the ground where it's fallen from an outhouse or privy that must be manually cleaned out. They walk through the villages carrying their neighbors' waste in containers on top of their heads and they empty the containers into the nearest moving water source, most often a river or stream.

This has led to the continued degradation not only of the untouchables but also of India's environment. The Ganges, known as Maa Ganga, which means Mother Ganges, is India's most sacred river. Lord Shiva is believed to have brought the river down from heaven in the locks of his hair. As a result of its divine beginnings, it is considered pure, and therefore all who plunge into her waters are made pure as well. Thus, hundreds of thousands of devout Hindus bathe in its waters every day and for added effect they drink the river's waters. Believers from thousands of miles away keep bottles of the river water in their homes, saving it for special religious events and even a bit of cooking.

And the Ganges is where nearly every Hindu wants his ashes scattered after a ritual cremation along the river's bank, ideally in the town of Varanasi, India's most holy city.

Yet the Ganges festers with untreated sewage and all those cremated bodies, many of which are only partially cremated. (The government released thirty thousand turtles into the river in a 1990s cleanup effort. The turtles were supposed to eat the uncremated body parts, but they all disappeared shortly after being released.) The fecal coliform count on the river near Varanasi is astronomical. Some checkpoints have recorded it as hundreds of thousands of times higher than acceptable levels. Tributaries to the river actually bubble with methane gas thanks to all the decomposing sewage.

Since India's scant wastewater treatment facilities were built nearly a hundred years ago by the British and were calibrated for much smaller populations, backed-up sewage overflows directly into the rivers. Also, the treatment plants do not have a decontamination stage, and partially treated water not only gets released to the waterways but is used by farmers for irrigation.

As a result, India's produce teems with bacteria and infectious diseases. The country has an infant mortality rate of sixty deaths in a thousand births and two million Indian children die every year of diseases due in part to poor sewage disposal.

Sewage is *the* scourge of India.

What can a country do when sewer systems are too costly and sanitation is at a breaking point?

It can turn to Dr. Bindeshawar Pathak, founder of the previously mentioned Sulabh International, who is building toilets to solve his country's problems.

"In my own country, how can anyone ignore the subject of the toilet when the society is faced with human excretions of the order of nine hundred million liters of urine and one hundred and thirty-five million kilograms of fecal matter per day with a totally inadequate system of its collection and disposal?" Dr. Pathak has asked. "Seeing this challenge, I think the subject of the toilet is as important as, if not more than, other social challenges like literacy, poverty, education, and employment. . . . The journey of the toilet has ended in Europe and North America but continues in the developing countries."

I had to see Dr. Pathak and his toilets: revolutionary, two-pit, pour-flush latrines. Together, they had helped free tens of thousands of untouchables, were cleaning the water supply, and were converting sewage into power.

• • •

What was the first thing I smelled upon disembarking from my plane at Indira Gandhi International Airport in Delhi, India's sprawling capital of thirteen million people? Urine. (I also thought I smelled it upon entering my hotel room later that night. The odor was probably either a figment of my addled imagination or the vapors from a common cleaning product.) Standing in the visa control line at midnight in eighty-five-degree heat, I failed to mention this to the two British-Indian women who were talking about how nice some children had been on the eight-hour flight from London's Heathrow Airport, but I did eventually tell them that I'd come to see a man about plumbing.

"You've come to India for plumbing?" one of them asked, bursting out in giggles. "There is no plumbing in India. That's why we live in England." Her companions choked with laughter. "Forget the toilets. Go climbing in the mountains with us."

I tried to inform them about Dr. Pathak's work, but they politely nodded and laughed some more.

"Just don't use a toilet outside your hotel," another warned as the line split apart. "It's not something you'll want to write about." There are roughly twenty five-star hotels in Delhi using a total of eight hundred thousand liters of water a day. The bathroom fixtures are like those you might find anywhere else in the world. Conversely, the majority of India's population have no bathrooms and get by on less than a liter of water per day.

Using a toilet in India, as well as in most other parts of Asia, the Middle East, much of Africa, and some parts of Europe is an odd experience for people used to a sit-down toilet, for the very fact that toilets in these places—outside of hotels, fancy restaurants, and museums—are of the squat type. It's very disconcerting the first time you try one out; you're constantly worried you might tip over and put your hand somewhere you don't want it to be. Also, since it is common practice to wipe with water in the Eastern countries, using the left hand, these toilets generally do not have toilet paper. That's why people with this custom do not eat with their left hands. I'll also hazard a guess that this is why we shake with our right hand; wiping with the left was once probably a more universal custom.

I arrived in Delhi in June, possibly the worst time as far as heat and dust go. It's a very brown time of year. While India might hold twenty percent of the world's population, it only gets four percent of the world's water supply, and most of that arrives during the monsoons, which weren't due for a few more weeks. Everything and everybody seemed to be waiting for the monsoons to wash all the dirt away. In the meantime, it was feet-burning hot. This I learned my first day there.

I was touring Delhi sites since it was a Sunday and I couldn't meet up with the Sulabh people until the next day. In one day, I visited nearly every tourist stop, from the president's palace, built by the British during the colonial era, to the Jami Masjid, the country's largest mosque. The Jami Masjid can accommodate twenty thousand worshippers and is an open-air affair. You have to take your shoes off in a mosque, even one without a roof, and so I hopped from burning red sandstone tile to burning red sandstone tile, admiring, more than anything, the heat. While there, I touched the footprint of Muhammad embedded in marble (it was a few days before I wondered how one might make an imprint in marble) and viewed one of his hairs as well, kept standing in a bit of wax under a glass vial decorated with silver. I had to give the Keeper of the Hair a donation, and after that every kid within a few miles was asking me for money.

For the record, I wasn't planning on shopping but was talked into it by the guide I'd hired at the hotel. We ended up at a Kashmir bazaar of sorts where I was beguiled into buying bedspreads, tablecloths, pashmina scarves, traditional silk Indian outfits for my family and me with the long tops that go down to the thighs, and other things we didn't need. The sellers wanted $2,200, I fought them down to $1,500 and the merchandise was probably worth $650, in retrospect. At least I didn't buy the rugs, although I was plied with three cups of Kashmir tea brimming with cardamom, while a rug salesman taught me how to tell the difference between cheap rugs and good ones. It's all in the thread count, two hundred per square inch versus six hundred, and the dye, chemical versus vegetable, which can be deciphered by the rug's sheen. If when you rub the rug in one direction it is shiny and rub it in the other it is dull, then the maker has used vegetable dyes.

I eventually escaped the bazaar, nearly running into a cow wandering down the middle of the street at four-thirty in the afternoon. Luckily, it was a Sunday and the traffic seemed careful not to hit him. Delhi has thousands of cows that wander its streets on a daily basis. They're often rounded up, brought to a shelter, and then let go again; they're not slaughtered, since cows are sacred to the Hindu. More animals appeared after that: monkeys squabbling under a banyan tree, horses pulling some merchant carts, and then there was an outdoor free animal hospital housing hundreds of tired-looking donkeys and small horses.

The next morning, after a breakfast of curried chick-peas and deep-fried potato patties, I headed off to Sulabh's Delhi headquarters. Wanting to be respectful, I wore the olive-green linen suit I got married in. This turned out to be one of Delhi's hottest days for the year: 45 degrees Celsius, or 113 degrees Fahrenheit. Most of the tour was conducted outside. Very quickly, I understood why Indians wore those flowing loose tops and billowy pants. However, I did not faint.

The Sulabh complex sits on Palam-Dabri Marg, a frenetic street in the Mahavir Enclave in southwest Delhi. Its low-slung cluster of buildings, painted a mellow pink and white, seemed to form a kind of fortress within the ever-growing, engulfing mass of humanity known as New Delhi. And the place was big: it housed a public school for untouchable children, a research station for improving India's sanitation, a vocational school for sewing, electrical work, computer repair, embroidery, typing, fashion designing, and beauty care, and a bevy of sanitation demonstrations from a methane-burning generator to nearly a dozen different outhouses displaying Dr. Pathak's two-pit, pour-flush latrines. A life-sized statue of a woman carrying a jerry can, presumably for human waste, stood at the entrance to the Courtyard of Latrines. She is an untouchable and a big red X was painted on the can.

It was hard to decide what to look at first, but before I could inspect anything, I was whisked into a dusky assembly hall. It was time for morning prayer.

"It is not a prayer for just one religion," Anita Jha, a

Sulabh vice chairperson in charge of vocational studies, whispered to me as we waited for the students to assemble. "But something for everyone." The elementary school students had left for summer vacation and so only the older vocational students were present, about seventy, most of them short untouchables in their late teens and early twenties. They were of a shorter stature than Indians of the upper classes because of malnourishment.

A student read a thought for the day and then walked over to where I was standing on the stage with some Sulabh officials. She draped a rose garland, a sandalwood garland, and a silk shawl over my shoulders. I turned redder than the flowers, whose smell was suddenly making me feel light-headed. Sweat was pouring down my face, tickling my back, soaking my underwear. But it wasn't the suit or the heat making me feel so odd. I was slowly realizing that this gathering was partially in my honor, and not only that, but I knew, all the way down to my now-curled-up toes, that I was going to have to say something—make some sort of formal comment. It's not every day a white guy shows up to study what Sulabh is doing, and things grew worse when I was introduced as a great sanitation scholar. Suddenly I was standing at the dais with nothing to say. So I said practically nothing, muttering something about how happy I was to be there. The students, appreciative of a short speech, awarded me with rousing applause.

Finally it was time for the prayer, which turned out to be a hymn. Everyone pressed their palms together in front of their chests in the traditional Hindu prayer style. The hymn was sung in Hindi, in a lilting, high-pitched tone. Later, I was given a translation. Dr. Pathak wrote the hymn, and all five stanzas end with the line: "Let's all come together and build a happy Sulabh world." There is no mention of toilets or improved plumbing.

After that, we went to the Courtyard of Latrines. It looked like the display grounds for miniature modular homes, but instead of houses there were rows of different outhouse models. Everything was well manicured and each outhouse had its own separate walkway. Some of the outhouses were covered with stucco. Others made of wood. And another was round, like a traditional rural home with a thatched roof. They all had one thing in common, though: the Sulabh two-pit, pour-flush latrine. But just what was a two-pit, pour-flush latrine? Why two pits? How was it different than a septic system? I was finally going to find out.

You enter the room like any outhouse—by opening the door. Then in a space about the size of an airline water closet, you're greeted with what they call the pan, or toilet bowl, set in the floor right in front of you. Made of fiberglass, concrete, or ceramic, it is shaped like a super-elongated toilet bowl and has a sloped bottom to drive everything toward the drain. The bottom of the pan is connected to a specially designed plastic question-mark-shaped water trap located below the floor that holds about a liter of water. And there's the first bit of Dr. Pathak's genius: a flush toilet that uses less than a quarter of a gallon for liquid waste and less than half a gallon for solids. There are no mechanical parts to break. You do your business, pour a liter of water from a handy pitcher straight into the pan, and it's gone. This is such a small amount of water that even places suffering water shortages are able to have flush toilets.

Then comes the second part of his genius: the two pits. What makes them so wonderful? Only one is used at a time, allowing waste in the unused pit to turn into beneficial compost. The trap is connected to a Y that leads to the two separate leach pits. A rock or some other form of obstruction blocks one arm of the Y. The open arm of the Y enters the pit, which is roughly one-meter tall by one-meter wide and is lined with bricks, rocks, or concrete in a manner that allows liquid waste to escape. Its bottom is unlined so that the dirt floor can facilitate decomposition. Both pits are covered with a concrete or metal lid.

It takes roughly three years for a pit to fill up when used by a family of five, and when one does, the obstruction in the Y is simply moved so the other pit can be filled. About a year and a half later, the waste in the first pit is completely broken down and safe for human handling. This natural decomposition is the key to Dr. Pathak's design. Typically, the emptying of the pit would be done by an untouchable on account of the wastes' hazards, but thanks to the simple science of this system, now anyone should feel comfortable picking the stuff up. There is no need for an untouchable—a caste-shattering result.

According to Hindu scriptures, God created four castes. The Brahmin, the highest caste, was created from God's head. The Kshatriya, made from God's arm, were the warrior class. The Vaishya came from his thighs and were the merchant class. The Shudra came from his feet and were the servant class. The untouchables, called Bhangi, a Sanskrit word meaning scavenger, did not belong to any of these castes and have thus always been considered outcasts. In the past and even to this day in some areas, the untouchables were not allowed to draw water from public wells and were not allowed to enter places of worship because their main job was to clean the dry latrines of the upper classes. (A dry latrine is anything from a bucket to an outhouse that is cleaned out in back.)

This caste system has existed in written form since 1400 BC and is based on the preexisting Aryan civilization that the Hindus came from. In other words, it's been a long time in the making.

Mahatma Gandhi was the first prominent Indian to promote untouchable liberation by not only calling for their liberation, but also requiring his followers to take care of their own waste when convening for political gatherings. While on his Salt March, he drank water from homes of the untouchables. He also proposed changing their name to Harijan, children of God (today, they refer to themselves as the *dalit,* a Sanskrit word meaning crushed and downtrodden). While his followers complied at the time, neither the new name nor the idea of saving the untouchables caught on. India did outlaw untouchability in 1950 under its post-independence constitution, but this was never enforced. According to recent figures, there are 160 million *dalits* and roughly a million of them still work as indentured servants, manually cleaning out latrines. They remain downtrodden. A 1989 law banning atrocities to *dalits* is revealing in what it prohibits. This law made it illegal to force them into bonded labor, to deny them access to public places, to foul their drinking water, to force them to eat obnoxious substances, or to parade them naked or with painted faces.

When India was celebrating the centennial of Gandhi's birth in 1968, Dr. Pathak joined the Bhangi-Mukti, Untouchable Liberation, wing of the celebration committee in his home state, Bihar. The more he learned of their plight, the more he became determined to do something. Eventually he went back to school and got a Ph.D., with his thesis being the study of scavenging (meaning the work of untouchables and nightsoil men). During his studies, he lived with untouchables and developed the two-pit system as a way to lift this entire class of people out of the sewers.

Dr. Pathak is a tall, dark-haired Brahmin—the upper crust of the Indian caste system—and his family and in-laws were more than a little shocked by his chosen calling. "Toilets were not something that we talked about in the Brahmin class and we especially did not make our living with them. My father-in-law wanted me to quit this business right away and get a real job—perhaps go back to teaching as I had been before," he told me. "I explained to him that I wasn't doing a job. Instead, I was making a history, something that we could look upon for guidance. I was creating a path that all of India could follow."

"Did that work—telling him that?" I asked.

"Well, soon after, different municipalities started placing orders for us to build our latrines and the movement started to become a financial success." So his father-in-law got over his objections.

While the majority of the toilets Sulabh has helped build have been for individual homes, the heart of the movement is in providing public facilities. Many of the people Dr. Pathak wants to reach with his facilities don't have homes or land. They live in "houses" made of vinyl tarps in areas that look more like refuse piles than neighborhoods.

"Everybody had laughed at us for having the idea that people would pay to go to the toilet," he continued. "But we were charging something less than one of your pennies. We had five hundred users the first day!" Dr. Pathak's organization went from being in a situation in which he had to manually help the mason move his cart through the streets so the work could be completed within budget, to what it is today, a company with more than one million latrines built, 119 biogas operations, and six thousand public all-in-one facilities that house showers, toilets, and medical aid, with an operating budget of $20 million.

"What was needed all along was to make sanitation a business—something that money could be made from.

And we had to make people drop the toilet taboo. When I started this, people would never talk about toilets or human excreta—it just wasn't allowed. Now everyone does, from the government on down. This has happened in just thirty years. It's a complete cultural change after thousands of years."

As if to illustrate this point, I'm shown a bag marked "human excreta" that is filled with dried, composted human waste that has been dug out of one of their pits. They encourage visitors to pick up a handful and smell it. It's obviously a dare, but an important one if taboos are going to be broken.

I scooped up a handful with my left hand. It had no odor. "Nice," I said, as if commenting to a waiter after testing the wine. Since I don't know what I'm looking for in either case, it's an easy thing to say.

"This can now be used as fertilizer," a scientist explained. "When we first started, we noticed that it was not very porous and had a difficult time mixing in with the soil, so we developed a machine to granulate the manure." He then showed me a container of human waste that looked more like sand. Later, I saw a lovely flower garden that was fertilized with this waste.

Ah, but how are the flower garden and lawn staying hydrated, I asked, given Delhi's withering early summer heat? And that was where things grew even more interesting. The garden was watered with organically treated wastewater. I was starting to giggle, things were getting so cool—figuratively, that is.

The lawn and flower garden were located directly behind public toilets that line the front wall of the Sulabh complex. (These toilets, by the way, were extremely clean. I walked in unannounced and all of the separate stalls were immaculate. Two boys were crouched under a shower head, washing up for the day. Someone else was doing wash.) The waste from these facilities was collected in a large steel drum buried beneath the garden, which served as a digester to make biogas—a methane-rich substance that should soon be powering the world. Biogas can be burned by gas cooking stoves, water heaters, and even generators. Practically anything run on natural gas can be run on biogas. And the supply is endless.

Dr. Pathak first got the idea of building small-scale biogas plants when he overheard a conversation about a man powering things at his farm by getting gas from his and his animals' waste. He interrupted the two men talking at a restaurant table near his and asked where the man lived. Eventually, he tracked the biogas man down, and after studying his system Dr. Pathak decided it could be done with human waste. Sulabh built its first biogas plant about a decade ago. Interestingly, the first recorded biogas plant in the world was developed at a leper colony in India in 1859 but wasn't duplicated elsewhere until another was attempted in Exeter, England, in 1895, where the captured gas was used to fuel street lamps. The Exeter plant used gas accumulated from sewage and was thus perhaps the first example of Dr. Pathak's kind of system. Other European municipalities sporadically contrived similar systems throughout the twentieth century, but the technology and knowledge of these experiments was not disseminated around the world. They were isolated events and so Dr. Pathak has created his system virtually from scratch.

It took some head-scratching trial and error at first, but eventually they came up with a simple, nearly fail-safe operation. When a digester—the first stage of the biogas plant—first comes online, cow dung and water hyacinth are added to get microbes fired up, but eventually all it takes to produce biogas is unadulterated human waste. Here at Sulabh headquarters, a vent pipe from the digester traps gas that is then used for a stove (on which my lunch was cooked that day), gas lanterns, and a water heater. To make this happen, people simply have to use the toilets and the digester needs to get cleaned out periodically. The digested solid waste, after being dried out, is used as compost. The liquids are drawn into another storage tank from which they are pumped into a charcoal/sand filter. It then flows into still another tank, where it is treated with ultraviolet light.

While my hosts were telling me this, they were pointing out the various tanks, pumps, and filters. At the end, we came to a tap coming off the ultraviolet tank. It was turned on and a bottle of water was collected and then held before me. I wasn't sure if I was supposed to drink it or what, so I just smiled. The tank attendant motioned me closer and then stuck it in my face. I hesitated and then realized, finally, that I was only supposed to look and smell. It was odorless and crystal clear and, I then learned, was used to water the lawn and flowers. That's how everything was staying so green and colorful—from human waste.

"This is a much better system than the sewer system," Dr. Pathak told me. "I do not want to criticize the past, but it is true. The biogas system, and even the two-pit system, does not pollute the water supply and it captures otherwise wasted energy. In the future, human excreta must be taken care of on location like in a biogas digester. The remaining gray water from showers and sinks, though, can be sent on via the sewer system to the treatment plants. I believe this should be the future."

• • •

I'd gone to India to learn what they were doing to alleviate their sewage problems but instead found a solution for our own. In Dr. Pathak's world, what we spend billions of dollars to "purify" through chemicals is turned to energy and clean water for relatively little money. He has brought a future solution to the present.

Dr. Pathak, however, does not see his work as complete. His organization has liberated sixty thousand untouchables. "But, although the government does not say this, more and more dry buckets are being used as India speeds into the future," he worried. "The government says there are only six million of these latrines in existence, but our figures make it closer to ten million. The politicians say they want to do something about this, but what they say and what they do . . . there is no connection. They do not follow up. In Delhi alone, we have four million people going outside for defecation. They live in squatters' shacks and have nowhere to go. But the politicians won't do anything. So, we are building the toilets."

According to Dr. Pathak's estimates he still needs to build another 140 million toilets at an estimated 6,000 rupees per toilet, totaling roughly $19 billion. And he thinks he can achieve this within the next twenty years, if his government will get a little more committed. "Maybe those in power should remember how their relatives in the country must use the toilet—their own grandmothers and uncles. Maybe then they would remember to do something."

Maybe they should also remember the hope not only of the revered Mahatma Gandhi for the *dalit* but also of Nehru, India's first post-independence prime minister, for his entire country: "The day every one of us gets a toilet to use, I shall know that our country has reached the pinnacle of progress."

12

The Power of Poop

IN AMERICA and many countries around the world, we've gone from toilets that use 5 to 6 gallons per flush to 1.6 gallons. Low-consumption toilets obviously cut down on water usage, but they also make things easier on the other end. Less volume at the treatment plant means less chemicals and less energy wasted in making sewage safe for the environment.

Last Christmas, Santa Claus brought my family a toilet Dr. Pathak would be impressed with: a Caroma dual-flush Caravelle.

The Caravelle, made in Australia, where saving water is a national cause, has two flushing options. One button uses 1.6 gallons and the other just .8 gallons. And, unlike a lot of other low-consumption toilets on the market, the Caravelle, with its humongous four-inch trapway, is virtually impossible to clog. My son Angus loves to stuff entire rolls of toilet paper in our toilets. The Caravelle is the only toilet that doesn't miss a flush. "I know I sound like a used car salesman, but it's really true. These things never need plunging," John Karas, business development manager for Caroma in the U.S., told me. "I've had them in my home for the last two years and have never plunged, even when we have parties. I have a distributor who has eleven kids and he's never had to plunge them. The kids simply can't block them up. The only time I've ever heard of one getting stopped up was in a school in Texas where somebody stuffed an entire towel in the trap."

Is the Caravelle the future? It, and all the other 1.6-gallon-per-flushers out there, is a step in the right direction, but is it a big enough step?

The one thing I've learned visiting Dr. Pathak, strolling through London's sewers and watching plumbers work, is that plumbing has to change. Plumbing and sewerage are mired in the past and it's time for some entirely new thinking. With this in mind, I went in search of the future both online and at ISH North America 2004—"the largest consolidated trade show in North America for

Kitchen and Bath, Plumbing, PVF, Heating and Air Conditioning." And at first I couldn't find it. I saw some cool things but nothing that was going to change the world.

At ISH, housed in the Boston Convention and Exhibition Center—a space-age glass-sided construct that looks like an oversized airport terminal wearing a wide-brimmed hat—I spent an entire day looking and asking for *the* thing that was going to redirect the plumbing world. I saw a few novel products like copper pipes that could be pushed together instead of soldered and a kit that could convert 6-gallon toilets to 1.6-gallon water conservers. Mostly, though, it was endless isles crammed with display booths pushing the best wrench, plastic pipe, or marketing plan

"Um, there's a waterless urinal being talked about," a toilet-maker rep told me, registering the look of despair on my face. "I haven't seen it, but that's gotta be pretty good, although I bet it's pretty messy."

I didn't find the elusive waterless urinal at the convention but did come across it later on the Internet. A manufacturer aptly named the Waterless Company has actually been making its urinal since 1991 and has sold its product worldwide. (Even the Jimmy Carter Library has them.) Having such a urinal can save up to forty-five thousand gallons a day in commercial usage, and at home, if you have one boy under the age of thirteen, I figure it can save twenty-five hundred gallons a year (based on the boy peeing seven times a day and making it to the toilet/urinal five of those times).

How it works is pretty simple. A replaceable plastic trap sits in the drain of a urine-repellent bowl that can be cleaned with regular detergents but requires no water for flushing. The trap allows urine to enter along the rim, and the urine then flows through a chemical layer called BlueSeal. It then flows up and over into the drain as more liquid—more urine—is added. The BlueSeal is good for about fifteen hundred urinatings, and then another three ounces are poured in. It's biodegradable and therefore safe for the environment. "BlueSeal has a specific gravity that is lower than urine—similar to oil floating on water," an instructional video proclaims. The urine and its odor sink below the BlueSeal.

The trap, a patented piece of plastic called an Ecotrap, has to be replaced three to four times a year, depending on traffic. The plastic trap sits inside the base of the urinal and is cylindrical. You urinate onto the top of the trap, which is sealed except on the edges, where there are slots for the urine to pass through. The company sells a device called an X-traptor for removing the trap, since it's embedded in the drain in such a way as to deter vandals. (Why a vandal would have his hands down the urinal drain is anybody's guess, but then I'm usually confused by the genius of vandals.)

Okay, but how do you clean such a urinal? The company's literature says it should be cleaned as often as a regular urinal. That's fine, but how does it stay waterless in this situation? You've got to wash the cleaner down for appearance's sake, but of course if it's a man doing the cleaning, he can just urinate onto the bowl. The instructional video does not suggest this, however, and it cautions against pouring large amounts of water into the urinal, as this would flush away the BlueSeal.*

The Kalahari, Waterless's best-selling urinal, has a comfortable curvaceous design. It actually looks like you could sit on it, which brings up the issue of women's urine. The men get to conserve water but the women don't?

These urinals have limited appeal for other reasons as well. Most people don't have urinals in their homes and the urinals don't do anything for solid waste.

No, what's needed is something that truly transforms

* Another company, Falcon Waterfree, has a similar item on the market and was founded by one of the original Waterless owners. The Falcon trap and urinal cost a little more but appear to have superior design because the chemical liquid barrier does not go down into the drain as it does with the Waterless system. You do have to replace the trap after seven thousand uses, though, as a buildup of uric acids, and other small items like hair, will eventually clog the trap.

the world of plumbing. Something that changes the way we handle both supply and waste forever. Something that I was beginning to worry may not yet be on the horizon in the Western world—until my father, ever mindful of my scatological interests, sent me a news clipping touting just such an invention. In truth, he actually sent me a clipping about a book that had nothing to do with innovative plumbing, but as luck would have it, an accompanying article touted this breakthrough. What exactly is this revolutionary concept? Sewage-powered fuel cells.

A fuel cell is an electrochemical energy conversion device—in other words, a type of battery. Most of us have heard of fuel cells that run on hydrogen and oxygen. They're going to be the next Big Thing in automobile propulsion. At least, that's what we're told. The way they work is that the cell converts these two components into water. This process creates electricity. As long as you keep feeding the cell more hydrogen and oxygen and let the water flow out of the cell, you'll never run out of electricity. A regular battery, on the other hand, stores chemicals inside a closed cell; their interaction creates electricity. After a while, they don't react to each other and the battery goes dead. So the beauty of the fuel cell is that it's an endless battery.

However, the big drawback to fuel cells is that compressed hydrogen is difficult to come by. (The oxygen component is easy; existent fuel cells can pull it straight from the atmosphere.) Researchers have developed a device called a reformer that converts alcohol into hydrogen, but it's poor-quality hydrogen and emits gases other than hydrogen. This has led scientists to find other fuel sources, from natural gas to methanol. Or, in the case of a team of environmental engineers at Pennsylvania State University: wastewater. When the bacteria in the microbial fuel cell (MFC), as it's called, metabolize the organic matter in the wastewater, electrons are released, creating a usable supply of electricity.

The current Penn State fuel cell is a tiny Plexiglas cylinder, six inches tall by two and a half inches in diameter. Eight pieces of graphite, serving as anodes, sit in the main chamber, surrounding the cathode. The cathode—which is a carbon/platinum catalyst/proton exchange membrane or, in more simple terms, a type of carbon paper—is attached to a hollow tube within the cylinder and a wire connects the anode and cathode. Wastewater, taken from the top of a settling tank, is pumped into the chamber. Bacterial digestion of the wastewater releases electrons into the wire to the cathode. The process also releases hydrogen ions, which eventually limit the oxygen demand in the wastewater, which allows microbes to go even wilder in digesting the bacteria. Meanwhile, oxygen reaches the cathode from the air, as do the hydrogen and the newly developed electrons—all of which combine to create fairly pure water (seventy-eight percent of organic matter is removed) from the wastewater. This model MFC creates enough energy to run a small fan and can do so on just five and a half ounces of typical wastewater. Dr. Bruce Logan, head of the group creating the sewage-powered fuel cell, believes he's found the future of wastewater treatment.

"Even in places fortunate enough to have wastewater treatment plants, there's little incentive to fix the plant when it breaks," Logan said. "It's just too expensive to run. But if the treatment plant also generates electricity, then it's viewed as a moneymaker and there's a great incentive to keep it running. If we can increase the power generation and decrease the cost of the microbial fuel cell, we can make clean water more available for both developing and industrialized nations."

Built on a bigger scale, an enlarged version of Logan's current microbial fuel cell could power fifteen hundred homes, but it would take the wastewater of a hundred thousand people to do so. This is a vast improvement over Penn State's earlier prototype, which couldn't even generate enough power to light a single Christmas tree light, yet it's far to go before we can power an entire household off its occupants' waste—but not too far a way to aim for.

If mismanaged municipalities in the U.S. and poor countries around the world are going to clean up their wastewater, this is where they must go. Our current best solution—burning biogas to power turbines or produce heat—works well but hasn't universally caught on. Perhaps the microbial fuel cell will. And just imagine the sense of accomplishment you'll have every time you sit on the toilet. Maybe it will even no longer be such a shameful act.

Conclusion

Do It for Your Plumber

I STARTED LOOKING into plumbing because I literally stumbled upon its beauty ten years ago in my basement. It seemed taboo, unheralded, and mysterious—the "dark art" of the housing industry, as Richard Trethewey says. It's turned out to be all that and more.

We have the right to ignore plumbing as we have done for centuries, but as can be seen in these pages, there's a huge cost. Water gets polluted, children die, and entire nations live in the filth of their own waste. Cities' water systems are patched together in piecemeal fashion and we get stuck in a game of never-ending catch-up. The harbor is polluted? We build a new treatment plant that takes care of the immediate problem—like in Boston—but because of a lack of interest and focus, the power invested in our waste mostly goes unused.

We find out lead pipes are bad for us and we switch to copper. We find out that leaching copper is perhaps just as unhealthy, and we . . . well, for the most part we do nothing, except switch to plastic. What happens when we find out what this plastic is doing to us? Jump to the next invention? Sometimes it's the only thing we can do.

Gunnar Baldwin, senior accounts manager for Toto USA, the maker of my Jasmin washlet, knows the sewer system is merely an outdated quick fix to the odors of the chamber pot—an out-of-sight, out-of-mind kind of operation that has degraded countless water environments. As a result, he's spent decades arguing with and cajoling regional, state, and federal agencies to make lasting, cohesive changes, all because he believes the sewage system was the most damaging invention in history to mankind.

As the world's largest toilet manufacturer, Toto is the epitome of the modern plumbing company and by its very nature relies on the sewer system as it is. So why is one of its representatives knocking modern plumbing?

"Toilets have been my cause for thirty years," he told me with a small chuckle when we met in the conference room at the Einhorn, Yaffee architectural firm in

Boston, where he was giving a talk entitled "The Greening of the Water Closet." Baldwin, dressed in a blue blazer highlighted by a snappy green bow tie, beamed with enthusiasm for his subject, but the audience, all attending to receive a mandatory continuing education credit, were less than enthusiastic. Why was he so fired up?

Back in 1975, Baldwin had remarried and found himself in a *Brady Bunch* situation, plus two—ten people in one house. Problem was, unlike Mike and Carol, Gunnar and his wife Heather didn't have Alice the housekeeper, and they weren't hooked up to a sewer system. "To put it mildly, the septic system was failing. It was nearly a crisis situation and we had to do something. I did some research and found this company called Microphor that had developed an air-operated toilet that used compressed air to force waste out of the bowl. They were mainly being used on trains. I bought a few, switched out my dinosaurs, and the septic system was saved."

Sensing both a calling and a new way to earn a living, Baldwin became a distributor for Microphor on the East Coast and has been, as he says, "selling things to reduce the wastewater stream ever since."

Microphor had a toilet that would only use half a gallon per flush, and wanting to sell more of the toilets he'd fallen in love with, Gunnar began advocating for governmental incentives to lower toilet water consumption. Gunnar's father was a conservationist and Gunnar envisioned his toilet work as picking up that mantle. Cut down on water waste and we'd be headed one step in the right direction.

Gunnar started writing to nearly every state environmental agency on the East Coast—from Georgia on up—to get them to propose changes in their regulations that would encourage low-consumption toilets, pointing out that by reducing outflow by forty percent, they would be garnering huge savings for local and state governments. He got a lot of polite letters in return, but he believes he planted a seed of change. At the time, toilets used five to eight gallons per flush and so cutting back to toilets that used two gallons or less would result in huge water savings and thus money as well. The EPA and environmental activists were pushing for reductions and one day, Gunnar figured, one of the agencies had to cave in. Water issues had become inescapable—especially in the closest big city to Gunnar, Boston.

Finally, Massachusetts changed its code in 1989, mandating that all new toilets placed in homes and businesses must use no more than 1.6 gallons per flush. (Why 1.6? Because the low consumption toilets developed in Sweden and other Scandinavian countries were only using 6 liters, or 1.6 gallons.) Thereafter, other states followed Massachusetts's lead and in 1992 1.6-gallons-per-flush toilets became the only national rule regarding plumbing, as part of a National Energy Act amendment. (There is no national agency overseeing plumbing.)

And it almost didn't happen.

"The night before the code was to go into effect," Gunnar told me, "The Massachusetts Plumbing Board invited everyone from the industry to give testimony. Everyone had been given two years to develop this technology, but a representative from a major American manufacturer shows up claiming the whole thing was impossible. That it was going to be a disaster. That all the drain lines were going to be clogged. That there hadn't been enough time for testing. They couldn't possibly get a product out. It was ridiculous."

According to Gunnar, the chairman of the Massachusetts Plumbing Board—a regulatory agency—responded, "You should be ashamed. You've had years to come up with this."

Seeing that his whining tactic hadn't worked, the rep, according to Gunnar, admitted that this manufacturer did actually have a 1.6-gallon-per-flush toilet that would be available the next day. Of course, as many who bought its early 1.6-gallon-per-flush toilets know, they had a design, all right. It just wasn't a good one.

At first, what with poorly designed toilets and rampant toilet-running from Canada, people weren't switching in great volume to the low-consumption toilets. There was not, and still isn't, a law saying you had to replace your existing toilet, and if it couldn't flush your waste as well as the old ones, why switch? Then came the financial incentives. The Rocky Mountain Institute produced a report showing how much water and money Denver could save if it replaced its existing toilets. Seeing this, New York City's Department of Environmental Protection instituted a toilet rebate program in 1994, offering $240 to change out old water-guzzling toilets, some of which used up to eight gallons a flush. New York City ended up replacing 1.3 million toilets and studies showed an average of thirty-seven percent savings in water consumption per building. As Gunnar said, "This was the final nail in the five-gpf toilet's coffin."

After his talk concluded, where the only question asked concerned an architect's malfunctioning Toto toilet, Gunnar remained enthusiastic.

"The first inventors of the S-trap were simply trying to reduce the smell commonly endured in even the most elegant castles before the servants emptied their lords' chamber pots," Gunnar told me later. "But that trapway had to be connected to a pipe. This pipe, in the beginning, went out to a ditch flowing by. When the ditch stank too much it was extended to a bigger ditch. And so it grew to reach the ocean or large lake into which it emptied: a wretched cocktail of whatever was added along the way. The primary purpose of sewers, until the first treatment plant was built, was to make the foulness disappear."

As we know, the results have been horrendous. Look all over the world, from Boston Harbor to the Thames River to Tokyo Bay and to the Black Sea—the last of which is fed by rivers loaded with sewage waste from seventeen different European countries. Each of the bodies of water has been wasted by pollution, much of it either untreated, or poorly treated, sewage.

Then where lies the future? In waste separation and an all-out commitment to biogas, or the fledgling sewage fuel cells being developed at Penn State? Perhaps. One thing is clear, though: more must be done than simply cleaning our waste at municipal water treatment plants and dumping the remaining effluents into our oceans, rivers, and lakes. There needs to be a next step. Right now we're polluting our environment with either untreated or poorly treated wastewater or with the by-products of cleaning our wastewater or, even when that is done well, with the wasting of biogas.

Is this really how we want to continue?

It's time for you to get up off your toilet and say you're not going to take it anymore.

Bash through the drywall behind your water tank, see how it's all put together, and then follow your pipes to their terminus. Stand at the chain-link fence separating you and your waste at the local sewage plant, breathe in those heady aromas, and decide it's time for a change. If you won't do this for your own sake, do it for your plumber. He's saved civilization many times in the past, but now he needs your help.

Postscript

Russell and Jasmin: Uncensored

I CAN'T HELP MYSELF. Something must be wrong. "You idiot. Did you figure out the controls?" I ask, unable to hide my frustration.

"Yeah . . . What are you doing here?"

"Nothing," I lie. What's with this guy? Where are the exclamations? No one can resist Jasmin. Personally, I sit on her even when I don't have to use the bathroom. It's one of the best experiences *I've* ever had.

Then, I hear the dryer come on. Russell's only used the wand for twenty seconds at the most. That's not enough time. He's not doing it right.

"Is that the dryer?"

"Yeah."

"Already?"

"Yeah."

"That's too soon. The wand won't have done its thing. You're not going to be clean enough."

"You think?"

For some reason, I take delight in his panicked tone.

"Oh yeah, definitely. It can take a few minutes. Try the oscillating button. It makes the water go everywhere . . . Did you try high pressure?" My girls had used the toilet last, and they like the water to come out very gently. The first time the twins tried it they both said, "I don't like it. That's weird." But they've returned again and again. When it was Helen's first try, she jumped just a bit, then settled back down with a big grin on her face. Her favorite trick is to get it going and hop off so that it sprays whoever else might be standing in the bathroom.

"Yeah, I turned up the pressure," Russell says. Then he flushes—something the girls and I forgot about the first couple of times in all the excitement. (Lisa uses the washlet as well but only when the rest of us are not around.)

Russell opens the door.

I peek in and of course there's no smell at all. A fan has sucked air into the deodorizer and through the catalizer, which is coated with manganese oxide. Ionically charged oxygen atoms on the manganese oxide coating have actually pulled the stinky molecules apart. Afterward, they re-formed as less odorous molecules, which were exhausted back into the room. There are no filters to be replaced.

"Well, what did you think?" I have to ask.

"It's a video game for your butt," he answers.

"But did you like it?"

"It takes too long. I'm a fast wiper."

"How about efficiency? Did it get it all?"

"Oh, I should check?"

"You didn't follow up with any paper? You've got to use the paper. Everyone knows you've got to use the paper. Isn't your butt still wet?" I ask. The blowdryer, although plenty hot at 140 degrees, takes about four minutes to completely dry you—longer than most people are willing to wait.

He goes back in, closes the door. A few seconds pass.

"Oh, I don't know. Not bad."

"But?"

"Not bad, I'd say pretty efficient, all in all."

"How does it compare to the normal experience?"

"I got it done with one wipe but that's pretty efficient."

"I'll say," I say. Talk about understatement. What does he think I'm doing, recording his every word?

"On really bad days, you just give up, right. With this thing, you know you're walking away clean," I add, feeling I must pipe in.

He opens the door.

"Where does it exhaust to?" he asks peering over the seat. I tell him it has a built-in deodorizer that doesn't use filters. The virtually odor-free air exhausts right into the bathroom. I walk into the bathroom with him. It smells great—like the soap Lisa used in the shower the night before. "Pretty high tech," he allows.

"What did you think about the warm seat?" I ask.

"That's the best part," he says, finally smiling. It's like somebody's just sat there for you. Perfect."

"Didn't you like the warm water? It's 104 degrees."

He shrugs his shoulders and says, "It was kinda hard."

I shake my head and don't even bother reminding him that he could've used the "soft wash" feature.

He looks down. "Hey, what's this all over my pants?" The bottom couple of inches of his left pants leg is soaked. He looks worried.

"You must have peed on it," I offer, helpfully.

He rubs his pants leg. Smells his fingers. "Nope, no way. This thing's leaked on me."

I get down on the floor and inspect the base. There's water all over the place! "You did something wrong.

This has never happened. If you've broken it . . ."

"I did what you told me to do."

I sit down and look at the control panel. "You've got it spraying all the way forward, but you used the rear wash button, right?"

He nods his head.

"How could you have washed with it set so far forward? That's for people with much smaller bottoms."

"It seemed like the right thing to do," he answers, truculently.

I pull down my pants and give it a try. Water flies out, hitting the back of my legs. "Didn't you notice this happening when you were using it?"

"I thought that was what it was supposed to do." Huh?

"Well, it's not Jasmin's fault. You had it set wrong. You can't have it this far forward when you're as big as us. It doesn't get anywhere near the right place. That setting is for kids and little people and . . ." I'm feeling a little defensive for Jasmin, an early stage in our relationship that will eventually ease into a more comfortable understanding. For now, though, I've got to explain away this embarrassing problem as something wrong with Russell. "And, well, this toilet I've put it on is really too small. That wouldn't have happened on a regular-sized toilet."

"Sure," Russell says, heading out the room, back downstairs.

Lisa stops him on the way out. "What did you think?" she asks.

"It was fine," he says, "but my pants are soaked."

It's been months now and he hasn't returned to use my toilet.

I guess some people aren't ready for the future.

Selected Bibliography

Bourke, John G. *The Portable Scatalog.* New York: William Morrow and Company, 1994.

Fenichell, Stephen. *Plastic: The Making of a Synthetic Century.* New York: Harper Business, 1996.

Harington, Sir John. *A New Discourse of a Stale Subject, Called the Metamorphosis of Ajax.* New York: Columbia University Press, 1962.

Hart-Davis, Adam. *Thunder, Flush and Thomas Crapper.* North Pomfret, Vermont: Trafalgar Square Publishing, 1997.

Hodge, A. Trevor. *Roman Aqueducts and Water Supply.* London: Duckworth, 2002.

Horan, Julie. *The Porcelain God.* Secaucus, New York: Carol Publishing Group, 1997.

Lambton, Lucinda. *Temples of Convenience.* New York: St. Martin's Press, 1978.

Laporte, Dominique. *History of Shit.* Cambridge, Mass.: The MIT Press, 1993.

Lewin, Ralph. *Merde.* New York: Random House, 1999.

OGLE, MAUREEN. *All the Modern Conveniences.* Baltimore: Johns Hopkins University Press, 1996.

PUDNEY, JOHN. *The Smallest Room.* New York: Hastings House, 1954.

ROBINS, F.W. *The Story of Water Supply.* London: Oxford University Press, 1946.

STILLE, ALEXANDER. *The Future of the Past.* New York: Picador, 2002.

WRIGHT, LAWRENCE. *Clean and Decent.* New York: The Viking Press, 1960.

Acknowledgments

A Big Thanks to: George Haselton and his crew at Rockport Mechanical, Richard Trethewey and *Ask This Old House,* Thames Water, A. Trevor Hodge, the Manoogs and their American Sanitary Plumbing Museum in Worcester, Massachusetts, the helpful folks at the American Antiquarian Society, Jonathan Yeo, Charlie Tyler and Pat Costigan at the Massachusetts Water Resources Authority, Richard Remson at the Rockport Foundry, everybody at Sulabh International, Rick Knowlton at Aqua Maine, Gunnar Baldwin at Toto, and Rebecca Farley, my fellow sewer enthusiast.

I thank my agent, Sally Wofford Girand, for her insight and commitment and my editor, Peter Borland, for turning my mess into something readable.

And I want to thank Lisa for humoring me at great cost.